FLORA OF TROPICAL EAST AFRICA

VIOLACEAE

C. Grey-Wilson

Annual or perennial herbs, shrubs or small trees. Leaves spirally arranged, rarely opposite or verticillate, simple, entire or toothed, rarely dissected; stipules present, small or foliaceous, the margin often ciliate or laciniate. Flowers actinomorphic or zygomorphic, solitary or in a simple or compound inflorescence, often thyrsoid, terminal or axillary, usually hermaphrodite, sometimes unisexual on separate plants. Sepals 5, free or united near the base, usually persistent. Petals 5, free, equal or unequal, the anterior one (lowermost in flower) sometimes spurred, imbricate, generally deciduous. Stamens 5, antisepalous, the lower pair (anterior) in zygomorphic flowers each with an appendage which projects into the spur and which secretes nectar; filaments free or united wholly or partly into a ring around the ovary; anthers introrse, usually with a prominent connective appendage, sometimes with thecal appendages also. Ovary sessile, ± ovoid, 1-locular, with (2–)3, 4 or 5 parietal placentas; style solitary, often thickened towards the stigma, which is generally undivided. Fruit a loculicidal capsule, generally splitting into 3 wide-spreading contractile valves, rarely a nut or berry. Seeds generally with ample endosperm, occasionally arillate.

A family of about 22 genera and some 900 species, confined mainly to the Old and New World tropics and subtropics, however the genus *Viola* is predominantly temperate in distribution.

Flowers actinomorphic, the anterior petal not spurred or saccate;
  shrubs or small trees with caducous stipules . . . . . .1. **Rinorea**
Flowers zygomorphic, the anterior petal (lowermost in flower)
    spurred or saccate; herbs, rarely shrubs, with persistent stipules:
  Anterior (lowermost) petal sessile, without a claw; leaf-lamina
      cordate, reniform, suborbicular or oval; seeds smooth .2. **Viola**
  Anterior (lowermost) petal with a prominent narrow claw; leaf-
      lamina linear, linear-lanceolate to elliptical, always cuneate at the base; seeds longitudinally striate or punctulate-
      striate . . . . .3. **Hybanthus**

1. RINOREA

Aubl., Hist. Pl. Guian. Fr. 1: 235, t. 93 (1775)

Alsodeia Thouars, Hist. Vég. Isles Austr. Afr.: 55 (1805)

Shrubs or small trees, rarely above 20 m. tall. Leaves spirally arranged, rarely opposite or verticillate, usually petiolate, entire or toothed; stipules small but enclosing the terminal bud, caducous. Inflorescence a simple raceme or a compound raceme or thyrse, rarely reduced to a solitary flower; pedicels articulate. Flowers actinomorphic, very rarely slightly zygomorphic. Sepals ± equal, margin often ciliate. Petals ± equal, spreading or reflexed. Stamens free or the filaments wholly or partially united into a tube, with or without a free margin; anthers free with a thin decurrent or non-decurrent prolongation of the connective (connective-appendage), and generally also with 2 free or ± united ventral or thecal appendages. Fruit a loculicidal capsule with 3 contractile valves, each containing 1–several seeds, rarely semisucculent and indehiscent.

A genus of about 340 species, found primarily in tropical forests of both the Northern and Southern Hemispheres.

Flowers in axillary spikes or racemes, sometimes virtually fasciculate (never terminal):
Filaments free or fused together only at the very base; connective-appendage equal in length to the anthers, 1–1.5 mm.
long, not or only partially decurrent:
Inflorescence a long slender raceme or ± spike, the rachis
always exceeding 25 mm. long   .   .   .   .   .   .   .11. *R. convallarioides*
Inflorescence a short condensed raceme, the flowers often appearing fasciculate, the rachis never more than 15 mm.
long, often far less:
Ovary and fruit glabrous; connective-appendage orbicular, much broader than the anthers, not decurrent   .12. *R. beniensis*
Ovary and fruit pubescent; connective-appendage oblong,
scarcely wider than the anthers, partially decurrent   .13. *R. squamosa*
Filaments united to form a tube; connective-appendage twice
as long as the anther, 2.5–4 mm. long, decurrent:
Pedicels and peduncle puberulous to glabrous or with scattered short hairs; bracts soon caducous; petals erect or
spreading, only tardily reflexed; ovary and fruit glabrous
14. *R. elliptica*
Pedicels and peduncle densely pubescent; bracts persistent;
petals quickly reflexed at anthesis; ovary and fruit sparsely to usually densely pubescent   .   .   .   .   .   .15. *R. angustifolia*
Flowers in a terminal cymose panicle, or thyrse (sometimes also
axillary in *R. ilicifolia* and *R. subintegrifolia*):
Leaf-margin spinose-serrate, spines 0.5–4 mm. long   .   .   .1.   *R. ilicifolia*
Leaf-margin not spinose:
Lamina dotted beneath with small dark brown glands   .   .3.   *R. welwitschii*
Lamina not gland-dotted beneath:
Young shoots and young petioles pubescent, rarely sparsely so:
Ovary and fruit densely to sparsely pubescent; inflores-

cence almost as broad as long; leaf-apex long-acuminate, 15–40 mm. long above the uppermost tooth
  9. *R. ferruginea*

Ovary and fruit glabrous; inflorescence slender, at least twice as long as broad; leaf-apex obtuse or shortly acuminate, not exceeding 10 mm. long above the uppermost tooth:
Fruit-capsule smooth; petals thick, obtuse, strongly recurved at the tip; sepals usually pubescent.   .4.   *R. brachypetala*

Fruit-capsule rugose or scurfy; petals thin, acute or subacute, not recurved at the tip; sepals usually glabrous:
Lamina-margin serrate to subentire; fruit-pericarp 1.5–2.5 mm. thick   .   .   .   .   .   .2.   *R. arborea*
Lamina-margin finely serrate-dentate; fruit-pericarp ± 0.5 mm. thick:
Rachis of inflorescence 1.5–3.2 cm. long; fruit finely rugose; lateral veins of leaf-lamina 6–9 pairs.   .   .   .   .   .10.   *R. tshingandaensis*

Rachis of inflorescence 5–12 cm. long; fruit scurfy; lateral veins of leaf-lamina 8–12 pairs
  8.   *R. dentata*
Young shoots and petioles glabrous:
Inflorescence a slender thyrse, not more than 12(–15) mm. wide; peduncles* 2–5 mm. long:
Sepals scarcely overlapping at anthesis; petals thin, not recurved at the tip; fruit-capsule rugose   .6.   *R. subintegrifolia*

Sepals markedly overlapping at anthesis; petals thick, recurved at the tip; fruit-capsule smooth 4.   *R. brachypetala*

Inflorescence a broader thyrse, 20 mm. wide or more; peduncles 5–50 mm. long:
Petals coriaceous, strongly recurved towards the tip; fruit-capsule smooth   .   .   .   .   .   .   .4.   *R. brachypetala*

Petals rather thin, not or only very slightly recurved at the tip; fruit-capsule rugose (except for *R. scheffleri*):
Flowers erect at anthesis; pericarp 1.5–2.5 mm. thick   .   .   .   .   .2.   *R. arborea*
Flowers nodding at anthesis; pericarp thin, not more than 1 mm. thick:
Petals 3–3.8 mm. long; peduncles not more than 6 mm. long   .   .   .   .   .6.   *R. subintegrifolia*

Petals 4.5–6 mm. long; peduncles 5–50 mm. long:
Rachis of inflorescence 1.5–3.2 cm. long; leaf-

---

*The primary axis of the inflorescence is referred to as the 'rachis' throughout the text, secondary branches, linking the rachis to the lateral groups of flowers are 'peduncles'.

lamina narrowly elliptic to elliptic-oblanceolate, 4.8–14.5 cm. long; bracts and
bracteoles persistent . . . . .10. *R. tshingan-*
daensis
Rachis of inflorescence 5–15 cm. long; leaf-
lamina broadly elliptic to oblong or oblong-oblanceolate, 10–29 cm. long;
bracts and bracteoles caducous:
Leaf-lamina broadly elliptic, with pro-
nounced serrate-dentate margin, veins
pale beneath . . . . . . .7. *R. scheffleri*
Leaf-lamina oblong to oblong-oblanceolate,
subentire or inconspicuously serrate-
crenate, veins dark beneath . . .5. *R. oblongifo-*
lia

1. R. ilicifolia *(Oliv.) Kuntze,* Rev. Gen. Pl. 1: 42 (1891); Reiche & Taub. in E. & P. Pf. III. 6: 329 (1895); Hiern, Cat. Afr. Pl. Welw. 1: 35 (1896); Brandt in E.J. 50, Suppl.: 412 (1914); V.E. 3(2): 549 (1921); Exell & Mendonça, C.F.A. 1: 70 (1937); F.P.N.A. 1: 628 (1948); T.T.C.L.: 645 (1949); F.P.S. 1: 63 (1950); I.T.U., ed. 2: 448 (1952); F.W.T.A., ed. 2, 1: 99, 101 (1954); N. Robson in F.Z. 1: 250 (1960); K.T.S.: 599 (1961); Tennant in K.B. 16: 409 (1963); Taton in F.C.B., Violaceae: 9, fig. 1 (1969); Grey-Wilson in K.B. 36: 116, fig. 6A–D (1981); Hamilton, Ug. For. Trees: 130 (1981). Type: Angola, Cuanza Norte, Pungo Andongo, Barrancos de Catete, 1070 m., *Welwitsch 889* (LISU, lecto., BM, isolecto.!)

Shrub or small tree to 5 m., often less; branches greenish when young, glabrous or minutely pubescent at first. Leaf-lamina coriaceous, elliptic-oblong to obovate, 7.5–24.5 cm. long, (2.5–)3–11 cm. wide, the base cuneate to ± rounded or cordate, the apex acute to acuminate, glabrous or rarely pubescent on the midrib beneath; margin spinose-serrate, the spines 0.5–4 mm. long; lateral veins 8–11 pairs (10–16 in Madagascar); petiole 0.1–4 cm. long, glabrous or finely pubescent. Inflorescence a terminal, rarely axillary, narrow thyrse, the rachis 7–16 cm. long, finely pubescent; flowers greenish white, greenish yellow or yellow, sometimes with a tinge of orange, somewhat foetid; bracts and bracteoles triangular-ovate, 1.5–2 mm. long, caducous; peduncle 2–5 mm. long (6–9 mm. in Madagascar), finely pubescent. Sepals ovate-elliptic, 2.5–4 mm. long, ribbed, margin ciliate. Petals thick, oblong-lanceolate, 4.5–5 mm. long, acute to subacute, slightly recurved at the tip. Stamens 2.8–3 mm. long, fused below to form a tube without a free margin; connective-appendage oblong, obtuse; thecal appendages usually 2. Ovary glabrous; style 2.6–2.8 mm. long. Capsule 3-lobed, 12–16 mm. long, rugose, glabrous.

var. **ilicifolia**

Leaf-lamina cuneate to ± rounded at base; petiole 0.6–4 cm. long. Fig. 1/1–3.

Uganda. Bunyoro District: Budongo, Sonso R., *Eggeling* 2303! & Rabongo Forest, Feb. 1964, *H.E. Brown* 2024!; Mengo District: Mawokota, Kisitu Forest, Jan. 1932, *Eggeling* 147!
Kenya. Kwale District: Muhaka Forest, Apr. 1977, *Gillett* 21069! & Mwachi, Sept. 1953, *Drummond & Hemsley,* 4262!; Lamu District: NE. of Witu, Feb. 1956, *Greenway & Rawlins* 8950!
Tanzania. Bukoba District: Bushenya, Apr. 1948, *J. Ford* 316!; Uzaramo District: Pugu Forest Reserve, Kisarawe Market, Feb. 1977, *Wingfield* 3766!; Rujifi District: Mafia I., Uranzi

to Kikuni, Aug. 1937, *Greenway* 5081!; Zanzibar I., near Haitajwa Hill, Dec. 1930, *Greenway* 2657!

DISTR.   U 2; K 7; T 1, 3, 4, 6, 7; Z; Guinée to Sudan, south to Angola and South Africa (Natal)
HAB.   Lowland and submontane evergreen forest, usually in shade; 0–1800 m.

SYN.   *Alsodeia ilicifolia* Oliv., F.T.A. 1: 108 (1868)
   Rinorea khutuensis Engl. in E.J. 28: 436 (1900); Brandt in E.J. 50, Suppl.: 412 (1914); Engl., V.E. 3(2): 549 (1921); T.T.C.L.: 645 (1949). Type: Tanzania, Morogoro District, Ukutu [Khutu Steppe], *Goetze* 117 (B, holo., BM, K, iso.!)
   R. angolensis Exell in J.B. 73, Suppl., Polypet. Addend.: 12 (1936); Exell & Mendonça, C.F.A. 1: 70 (1937). Type: Angola, Cuanza Norte, Dondo, *Gossweiler* 9759 (BM, holo.!, K!, LISJC, iso.)
   R. ilicifolia Engl. var. *khutuensis* (Engl.) Tennant in K.B. 16: 411 (1963)

var. **amplexicaulis** *Grey-Wilson* in K.B. 36: 118, fig. 6D (1981). Type: Tanzania, Kigoma District, Uvinza [Uvinsa], *Bullock* 3242 (K, holo.!, iso.!)

Leaf-lamina base cordate, amplexicaul; petiole 1–3 mm. long. Fig. 1/4.

TANZANIA.   Biharamulo District: Nyakahura, Nov. 1948, *J. Ford* 862!; Mwanza District: Maisome I., Aug. 1958, *Procter* 980!; Kigoma District: Uvinza, Aug. 1950, *Bullock* 3245!
DISTR.   T 1, 4, 6; not known elsewhere
HAB.   Evergreen forest; ± 1150 m.

NOTE.   The Madagascan plant has rather narrower leaves with a greater number of lateral veins, while the peduncles are noticeably longer. This plant is assigned to subsp. *spinosa* (Tul.) Grey-Wilson.

2. R. arborea *(Thouars) Baill.* in Bull. Soc. Linn. Paris 1: 583 (1886); Perrier in Mém. Inst. Sci. Madag., sér. B, 2: 329 (1949) & in Humbert, Fl. Madag., Fam. 139: 42 (1955); N. Robson in F.Z. 1: 252 (1960); K.T.S.: 598 (1961). Type: Madagascar, without locality, *du Petit-Thouars* (P, holo.)

Shrub or small tree to 9 m. tall, sometimes more; young branches glabrous or finely pubescent at first. Leaf-lamina coriaceous, elliptic to oblong or oblanceolate, 5.5–30 cm. long, 2.4–12 cm. wide, the base cuneate to ± subrounded, the apex subobtuse, acute or sometimes shortly acuminate, glabrous or pubescent on the midrib beneath; margin serrate to subentire; lateral veins 8–12 pairs, prominent; petiole (0.5–)1–4.6(–6) cm. long, glabrous, rarely pubescent. Inflorescence a stout terminal thyrse, very rarely lateral; rachis 7–16.5(–20) cm. long, finely pubescent; flowers yellow, greenish yellow, greenish white or white, erect at anthesis; bracts and bracteoles lanceolate to triangular, 2–3 mm. long, pubescent, persistent; peduncle 12–30 mm. long, finely pubescent; pedicels 1–3 mm. long, finely pubescent. Sepals ovate to oblong, 2–4 mm. long, faintly ribbed or smooth, usually finely pubescent, the margin ciliate. Petals ovate-lanceolate to oblong, 4–6 mm. long, acute to subobtuse, scarcely recurved at the tip, glabrous or pubescent on the exterior. Stamens 3.5–4 mm. long; filaments united into a tube, with or without a short, free, lobed margin; connective-appendage triangular-ovate, ± acute, not decurrent; thecal appendage solitary, very short or absent. Ovary glabrous; style ± 3 mm. long. Capsule woody, unlobed, 15–25 mm. long, rugose, glabrous; pericarp thick, 1.5–2.5 mm.

KENYA.   Kilifi District: Marafa, Nov. 1961, *Polhill & Paulo* 798! & Vipingo, Dec. 1953, *Verdcourt* 1077!; Lamu District: Witu, Gongoni Forest, Dec. 1936, *Mohamed Abdulla* in F.D. 3846!
TANZANIA.   Pangani District: Bushiri, Dec. 1950, *Faulkner* 741!; Kilosa District: Vigude, Oct. 1952, *Semsei* 989!; Morogoro District: Turiani, Nov. 1955, *Milne-Redhead & Taylor* 7411!; Zanzibar I., Apr. 1870, *Kirk*!
DISTR.   K 7; T 3, 6, 8; Z; Mozambique, Madagascar
HAB.   Lowland evergreen forest; 0–850 m.

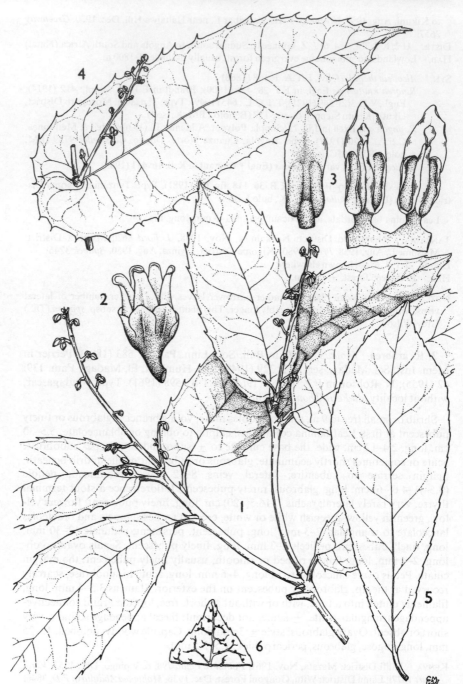

FIG. 1. *RINOREA ILICIFOLIA* var. *ILICIFOLIA* – **1**, habit, x ⅔; **2**, flower, x 4; **3**, stamens, anterior and posterior views, x 12. *R. ILICIFOLIA* subsp. *ILICIFOLIA* var. *AMPLEXICAULIS* – **4**, leaf and young inflorescence, x ⅔. *R. WELWITSCHII* subsp. *TANZANICA* – **5**, leaf above, x ⅔; **6**, leaf detail, lower surface, x 1⅓. 1, from *Brasnett* 359; 2, 3, from *Rawlins* 195; 4, from *Bullock* 3245; 5, 6, from *Semsei* 719. Drawn by Christine Grey-Wilson.

Syn.   *Alsodeia arborea* Thouars, Hist. Vég. Isles Austr. Afr.: 57 (1805)
       Rinorea orientalis Engl., V.E. 1(1): 290 (1910), *nomen nudum*

Note.   Specimens from Mozambique tend to have a leaf-margin which is markedly spinose-serrate, the teeth-spines up to 1 mm. long, though nowhere near the length of those on the leaves of *R. ilicifolia.*

3. **R. welwitschii** *(Oliv.) Kuntze,* Rev. Gen. Pl. 1: 42 (1891), pro parte; Brandt in E.J. 50, Suppl.: 416 (1914); De Wild. in B.J.B.B. 6: 146, 191 (1920); V.E. 3(2): 551 (1921); Melchior in E. & P. Pf., ed. 2, 21: 351, fig. 151F (1925); Exell & Mendonça, C.F.A. 1: 73 (1937); N. Robson in F.Z. 1: 251 (1960); F.F.N.R.: 264 (1962); Tennant in K.B. 16: 418 (1963); Taton in F.C.B., Violaceae: 59, fig. 22 (1969); Grey-Wilson in K.B. 36: 121 (1981). Type: Angola, Cuanza Norte, Golungo Alto, Mata de Quisuculo, *Welwitsch* 882 (LISU, lecto., BM, K, isolecto.!)

Shrub or small tree to 10 m. tall; branches densely to sparsely pubescent when young, eventually glabrescent, hairs generally rather bristly. Leaf-lamina obovate to oblanceolate, (5.5–)7–18(–21.5) cm. long, (1.5–)2.5–7(–10) cm. wide, the base ± cuneate, the apex abruptly acuminate, gland-dotted beneath, glabrous above, sparsely to densely pubescent beneath; margin crenate-serrate to serrate-spinulose, the spines short, up to 0.5 mm. long; lateral veins 7–12 pairs; petiole 0.5–3 cm. long, bristly pubescent. Inflorescence a slender terminal thyrse; rachis 4–12 cm. long, sparsely to densely pubescent; flowers yellow, ± pendulous; bracts and bracteoles lanceolate-triangular, 2–3 mm. long; peduncle (2–)6–13 mm. long, sparsely to densely pubescent; pedicels 1.5–4 mm. long, pubescent. Sepals ovate, 2–3.5 mm. long, puberulous to densely bristly pubescent, margin ciliate. Petals oblong, 4–5 mm. long, thick, obtuse, slightly recurved at the apex, glabrous to moderately pubescent outside. Stamens 2.8–3.2 mm. long; filaments fused into a tube for most of their length, with a free lobed margin; connective-appendage oblong to lanceolate, decurrent, almost 1.5 times the length of the anthers; thecal appendage short, entire or bilobed. Ovary glabrous to densely pubescent; style 2.5–3 mm. long. Capsule 3-lobed, 10–17 mm. long, smooth, coriaceous, glabrous or pubescent.

Syn.   *Alsodeia welwitschii* Oliv., F.T.A. 1: 110 (1868)

subsp. **tanzanica** Grey-Wilson in K.B. 36: 121, fig. 6E (1981). Type: Tanzania, Uzaramo District, Kiregese [Kurekese] Forest Reserve, Kisiju, *Semsei* 1380 (K, holo.!, EA, iso.)

Young branches and petioles sparsely pubescent to glabrescent. Leaf-lamina 7–16.3 cm. long, 2.5–6.5 cm. wide, pubescent usually only on the midrib beneath; margin generally serrate-spinulose; lateral veins 7–8 pairs. Inflorescence-rachis 4–5.5 cm. long, sparsely pubescent to puberulous. Sepals puberulous with a shortly ciliate margin. Ovary and capsule glabrous. Fig. 1/5, 6.

Tanzania.   Uzaramo District: Kiregese [Kurekese] Forest Reserve, Kisiju, Sept. 1953, *Semsei* 1380! & Vikindu Forest Reserve, Aug. 1953, *Semsei* 1301!; Lindi District: Rondo Plateau, Mchinjiri, Mar. 1952, *Semsei* 719!
Distr.   T 6, 8; not known elsewhere
Hab.   Lowland evergreen forest

Note.   This subspecies differs from subsp. *welwitschii* in being far less pubescent in all its parts and in having glabrous rather than pubescent ovaries and capsules. In subsp. *welwitschii,* which occurs in W. Africa, Angola, Zaire and N. Zambia, the young stems, petioles and inflorescences are often densely bristly hairy and the fruits are always pubescent to some degree. However, subsp. *welwitschii* is very variable and needs further investigation.

4. **R. brachypetala** *(Turcz.) Kuntze,* Rev. Gen. Pl. 1: 42 (1891); Th. Dur. & Schinz, Consp. Fl. Afr. 1(2): 209 (1898); Engl. & Brandt in Z.A.E. 2: 562 (1913); Brandt

in E.J. 50, Suppl.: 414 (1914); De Wild. in B.J.B.B. 6: 144, 153 (1920); V.E. 3(2): 550 (1921); Melchior in E. & P. Pf., ed. 2, 21: 351 (1925); Exell & Mendonça, C.F.A. 1: 69 (1937); F.W.T.A., ed. 2, 1: 101, 104 (1954); Tennant in K.B. 16: 421 (1963); Taton in F.C.B., Violaceae: 53 (1969); Hamilton, Ug. For. Trees: 130 (1981); Troupin, Fl. Rwanda 2: 432, fig. 136/2 (1983). Type: Zaire, Lower Congo R., *Christian Smith* (MW, holo., BM, K, iso.!)

Shrub or small tree to 6.5 m. tall, though often less; young branches pubescent or glabrous. Leaf-lamina obovate to oblong-elliptic or elliptic-oblanceolate, 6.5–19 cm. long, 3–8.5(–9) cm. wide, the base subrounded to ± cuneate, the apex acute or shortly and rather abruptly acuminate, glabrous or with some hairs on the midrib and lateral veins beneath; margin serrate-crenate to subentire; lateral veins 6–10 pairs; petiole 0.5–5.5 cm. long, glabrous or occasionally sparsely pubescent. Inflorescence a stout terminal thyrse, very occasionally one or two of the uppermost leaves producing a shorter axillary inflorescence; rachis 5–17 cm. long, pubescent; flowers cream, pale yellow or greenish yellow, nodding; bracts and bracteoles ovate to lanceolate, 1.5–3 mm. long, usually pubescent, margin ciliate, persistent; peduncle 3–17 mm. long, usually pubescent; pedicels 1–2 mm. long, usually pubescent. Sepals orbicular or elliptic, 2–3 mm. long, pubescent, rarely entirely glabrous, the margin ciliate. Petals oblong, 5–6 mm. long, obtuse, thick, the tip strongly recurved soon after anthesis. Stamens 3.4–3.8 mm. long; filaments* fused into a ring with a free margin, with or without a free portion to the filaments; connective-appendage ± ovate, 1.5 times the length of the anthers, decurrent; thecal appendages usually absent. Ovary glabrous; style 3–3.5 mm. long. Capsule 3-lobed, 11–16 mm. long, coriaceous, smooth, glabrous. Fig. 2/8, 9.

UGANDA. Acholi/Bunyoro District: Kabalega [Murchison] Falls, Feb. 1935, *Eggeling* 1601; Mengo District: Kyiwago [Kyewaga] Forest, Mar. 1923, *Maitland* 605! & Mpanga Forest, Oct. 1953, *Byabainazi* 69!
KENYA. Nandi District: Yala R., north bank, Apr. 1965, *Gillett* 16730!; N. Kavirondo District: Kakamega Forest, Jan. 1968, *Perdue & Kibuwa* 9472! & Mar. 1972, *Kokwaro* 3142!
TANZANIA. Bukoba District: Kaigi [Kaige], May 1935, *Gillman* 274! & 350!
DISTR. U 1, 2, 4; K 3, 5; T 1; Ivory Coast, Ghana, Nigeria, Cameroun, Central African Republic, Gabon, Angola, Zaire, S. Sudan, Zambia
HAB. Evergreen forest; 850–1900 m.

SYN. *Alsodeia brachypetala* Turcz. in Bull. Soc. Nat. Mosc. 36(1): 558 (1863); Oliv., F.T.A. 1: 109 (1868)
  A. aucuparia Oliv., F.T.A. 1: 109 (1868). Type: Angola, Cuanza Norte, Pungo Andongo, Barranco de Songue, *Welwitsch* 892 (LISU, syn., BM, K, isosyn.!)
  Rinorea aucuparia (Oliv.) Kuntze, Rev. Gen. Pl. 1: 42 (1891); Th. Dur. & Schinz, Consp. Fl. Afr. 1(2): 209 (1898); De Wild. in B.J.B.B. 6: 151 (1920); V.E. 3(2): 553 (1921); Exell & Mendonça, C.F.A. 1: 72 (1937)
  Alsodeia stuhlmannii Engl., P.O.A. C: 276 (1895). Type: Tanzania, Bukoba District, Kanjawassi stream, *Stuhlmann* 1626 (B, holo.†)
  Rinorea aucuparia (Oliv.) Kuntze var. *platyphylla* Hiern, Cat. Afr. Pl. Welw. 1: 35 (1896). Type: Angola, Cuanza Norte, Pungo Andongo, Mata de Pungo, *Welwitsch* 892b (LISU, lecto., BM, isolecto.!)
  R. congensis Engl. in De Wild. & Th. Dur., Ann. Mus. Congo, Bot., sér. 2, 1(2): 3 (1900). Type Zaire, *Dewèvre* 265 (BR, holo.)
  R. poggei Engl. in E.J. 33: 137 (1902); Brandt in E.J. 50, Suppl.: 414 (1914); De Wild. in B.J.B.B. 6: 143, 181 (1920); Engl., V.E. 3(2): 550 (1921); Melchior in E. & P. Pf., ed. 2, 21: 351 (1925); T.T.C.L.: 646 (1949); I.T.U., ed. 2: 448 (1952); F.W.T.A., ed. 2, 1: 101, 104 (1954); N. Robson in F.Z. 1: 251 (1960); K.T.S.: 601, fig. 110 (1961); F.F.N.R.: 264 (1962). Types: Zaire, Lulua R., *Pogge* 646 (B, syn.†) & Tondoa, under Kili, *Büttner* 492 (B, syn.†)

R. *stuhlmannii* (Engl.) Engl. in E.J. 33: 138 (1902); Brandt in E.J. 50, Suppl.: 414 (1914); T.T.C.L.: 646 (1949)

Alsodeia dawei Sprague in J.L.S. 37: 497 (1906). Type: Uganda, Toro District, Kibale Forest, *Dawe* 516 (K, holo.!)

A. poggei (Engl.) H. Dur., Syll. Fl. Congo: 35 (1909)

Rinorea dawei (Sprague) Brandt in E.J. 50, Suppl.: 418 (1914); De Wild. in B.J.B.B. 6: 159 (1920); Engl., V.E. 3(2): 553 (1921)

R. *moandensis* De Wild. in B.J.B.B. 6: 176 (1920). Type: Zaire, Moanda, 1907, *Gillet* 4011 (BR, holo.)

R. *pallidiviridis* De Wild. in B.J.B.B. 6: 179 (1920); Tisserant in Bull. Soc. Bot. Fr. 102: 35 (1955). Type: Zaire, Dima, Nov. 1903 *Laurent* (BR, holo.)

R. *seretii* De Wild. in B.J.B.B. 6: 185 (1920). Type: Zaire, Missa, *Seret* 270 (BR, holo.)

R. *verschuerenii* De Wild. in B.J.B.B. 6: 190 (1920). Type: Zaire, Yalala [Lalela], *Verschueren* 956 (BR, holo.)

NOTE. *R. brachypetala* is a widespread and variable species. The amount of pubescence on the inflorescence varies greatly from one specimen to another. The degree of fusion of the filaments is equally diverse; in some specimens the filaments are fused entirely, while in others only the basal third or half are fused together, leaving the upper part free.

5. R. oblongifolia *(C.H. Wright) Chipp* in K.B. 1923: 296 (1923); Melchior in E. & P. Pf., ed. 2, 21: 351 (1925); Exell & Mendonça, C.F.A. 1: 71 (1937); F.P.N.A. 1: 630 (1948); I.T.U., ed. 2: 448 (1952); F.W.T.A., ed. 2, 1: 101, 104 (1954); Taton in F.C.B., Violaceae: 27 (1969); Hamilton, Ug. For. Trees: 130 (1981). Type: Cameroun, Efulen, *Bates* 432 (K, holo.!, BM, iso.!)

Shrub or small tree to 15 m. tall; young branches glabrous. Leaf-lamina oblong to oblong-elliptic or oblong-oblanceolate, (7.5–)10–29 cm. long, (3.6–)5–12.5 cm. wide, the base subrounded to cuneate, the apex acuminate, glabrous on both surfaces, the margin shallowly serrate-crenate to subentire, lateral veins 6–12 pairs; petiole 1.5–5.2 cm. long, glabrous. Inflorescence a broad terminal thyrse, the rachis (5–)6.5–11(–15) cm. long, finely pubescent to puberulous; flowers yellow, orange or reddish; bracts and bracteoles ovate-lanceolate, 1.5–2.5 mm. long, soon caducous; peduncle 1.5–5 cm. long, finely pubescent to puberulous; pedicels 1–3 mm. long, finely pubescent to puberulous. Sepals ovate to suborbicular, 2.5–4 mm. long, often puberulous or finely pubescent, the margin ciliate. Petals oblong to ± oblanceolate, 4.5–6 mm. long, obtuse, scarcely recurved at the apex, usually puberulous or finely pubescent outside. Stamens 3.5–4.5 mm. long, the filaments fused for most of their length into a tube with a free lobed margin, glabrous to sparsely pubescent; connective-appendage ovate, decurrent, ± 1.5 times the length of the anthers; thecal appendages 2, linear-lanceolate. Ovary glabrous or puberulous; style 2.5–3 mm. long. Capsule 3-lobed, 15–22 mm. long, rugose, the surface often cracking into small 'scales'. Fig. 2/1–4.

UGANDA. Ankole District: Buhweju, road from Byabakara Camp, *Synnott* 38!; Mengo District: Kyiwaga [Kyewaga] Forest, Feb. 1950, *Dawkins* 509! & 24 km. Kampala to Mubende, Oct. 1932, *Eggeling* 537!

DISTR. U 2, 4; Guinée, Sierra Leone, Liberia, Ivory Coast, Ghana, Nigeria, Cameroun, Gabon, Zaire, S. Sudan

HAB. Evergreen forest; 1150–1450 m.

SYN. [*Alsodeia welwitschii* sensu Oliv., F.T.A. 1: 110 (1868), pro parte, *non* sensu stricto]
[*Rinorea welwitschii* sensu Kuntze, Rev. Gen. Pl. 1: 42 (1891), pro parte; De Wild. & Th. Dur., Ann. Mus. Congo, Bot., sér. 3(1): 12 (1901); De Wild. in B.J.B.B. 6: 191 (1920), *non* (Oliv.) Kuntze sensu stricto]
Pittosporum oblongifolium C.H. Wright in K.B. 1897: 243 (1897)

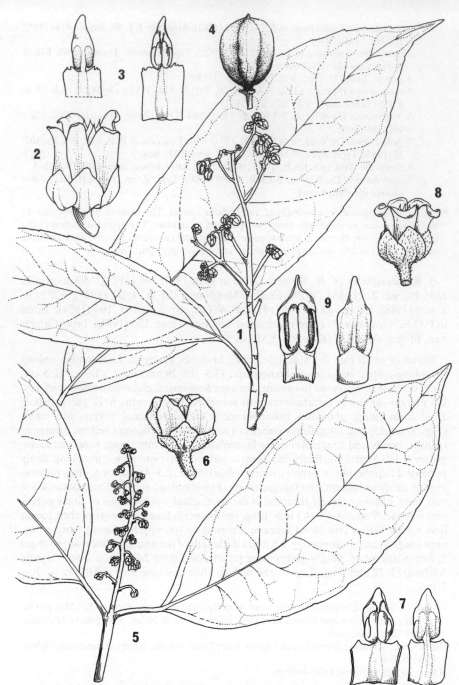

FIG. 2.   *RINOREA OBLONGIFOLIA* – **1**, habit, x $\frac{2}{3}$; **2**, flower, x 4; **3**, stamens, anterior and posterior views, x 6; **4**, fruit, x 1. *R. SUBINTEGRIFOLIA* – **5**, habit, x $\frac{2}{3}$; **6**, flower, x 4; **7**, stamens, x 8. *R. BRACHYPETALA* – **8**, flower, x 6; **9**, stamens, anterior and posterior views, x 8. 1–3, from *Dawkins* 509; 4, from *Dawe* 761; 5–7, from *Drummond & Hemsley* 3444; 8, 9, from *Gillman* 1935. Drawn by Christine Grey-Wilson.

6. R. subintegrifolia *(P. Beauv.) Kuntze,* Rev. Gen. Pl. 1: 42 (1891); Th. Dur. & Schinz, Consp. Fl. Afr. 1(2): 212 (1898); Brandt in E.J. 50, Suppl.: 418 (1914); De Wild. in B.J.B.B. 6: 149, 187 (1920); V.E. 3(2): 553 (1921); Pellegrin, Fl. Mayombe 1: 19 (1924); Melchior in E. & P. Pf., ed. 2, 21: 351, fig. 151H (1925); F.W.T.A., ed. 2, 1: 104, fig. 31/2a–f (1954); Tennant in K.B. 16: 430 (1963); Taton in F.C.B., Violaceae: 30 (1969). Type: Nigeria, Oware [Warri], *Palisot de Beauvois* (G, holo.)

Shrub or small tree to 4 m. tall, though generally less; branches usually glabrous. Leaf-lamina ± elliptic to elliptic-oblanceolate or oblong-obovate, (5–)7–26 cm. long, (1.5–)2.8–9.2(–10.7) cm. wide, the base cuneate to subrounded, the apex acuminate, acute to subobtuse, glabrous or occasionally with a few hairs on the midrib beneath, margin subentire to obscurely crenate-serrate, lateral veins 5–9 pairs; petiole 6–30 mm. long, glabrous. Inflorescence a slender terminal, occasionally lateral, thyrse, the rachis 4.5–7(–10) cm. long, usually puberulous or finely pubescent; flowers yellow, yellowish green or cream; bracts and bracteoles triangular-ovate, ± 1mm. long, eventually falling; peduncle 3–6 mm. long, pubescent or puberulous; pedicels 1–2 mm. long, pubescent or puberulous. Sepals ovate, 1.5–2.5 mm. long, finely pubescent usually, margin ciliate. Petals oval, 3–3.8 mm. long, pubescent or puberulous on the outside, margin ciliate. Stamens 3–3.25 mm. long, the filaments fused for a part or all of their length into a tube with a free, often rather uneven or toothed, margin; connective-appendage ovate-elliptic, almost twice as long as the anther, margin subentire; thecal appendages 1, subentire to bilobed, or 2. Ovary glabrous; style 1.7–2 mm. long. Capsule 3-lobed, 10–18 mm. long, the lobes broadly ridged, rugose, glabrous. Fig. 2/5–7.

TANZANIA. Lushoto District: E. Usambara Mts., Amani, July 1953, *Drummond & Hemsley* 3444!; Morogoro District: Turiani, Manyangu Forest, Nov. 1953, *Semsei* 1408!; Njombe District: Lupembe, R. Ruhudji [Ruhudje], 1931, *Schlieben* 1302!
DISTR. T 3, 6, 7; Guinée, Sierra Leone, Liberia, Ivory Coast, Ghana, Nigeria, Cameroun, Gabon, Zaire
HAB. Evergreen forest; 300–1800 m.

SYN. *Ceranthera subintegrifolia* P. Beauv., Fl. Owar. 2: 11, t. 66 (1808); DC., Prodr. 1: 314 (1824)
Alsodeia subintegrifolia (P. Beauv.) Oliv., F.T.A. 1: 109 (1868)
Rinorea amaniensis Brandt in E.J. 50, Suppl.: 418 (1914); De Wild. in B.J.B.B. 6: 149 (1920); Engl., V.E. 3(2): 553 (1921); Melchior in E. & P. Pf., ed. 2, 21: 351 (1925); T.T.C.L.: 646 (1949). Type: not cited

NOTE. *R. subintegrifolia* is very variable in leaf size and in the characters of the androecium. The proportion of free or fused filaments varies from plant to plant and even on the same inflorescence. *Greenway & Farquhar* 8638, from Tanzania, Morogoro District, S. Nguru Forest Reserve, has unusually long and narrow leaves, up to 26 cm. long and 7.2 cm. wide.

7. R. scheffleri *Engl.,* E.J. 33: 142 (1902); Brandt in E.J. 50, Suppl.: 412 (1914); De Wild. in B.J.B.B. 6: 142, 183 (1920); Engl., V.E. 3(2): 549 (1921); Melchior in E. & P. Pf., ed. 2, 21: 351 (1925); T.T.C.L.: 646 (1949); Grey-Wilson in K.B. 36: 120 (1981). Type: Tanzania, E. Usambara Mts., Ngwelo [Nguelo], *Scheffler* 26 (B, holo.†, K, iso.!)

Much branched shrub or small tree; branches glabrous. Leaf-lamina broadly elliptic, 12–23 cm. long, 4.6–10.5 cm. wide, the base broadly cuneate, the apex shortly acuminate, glabrous on both surfaces, margin finely serrate to serrate-dentate, lateral veins 8–14 pairs; petiole 3.5–5 cm. long, glabrous. Inflorescence a narrow pyramidal terminal thyrse, the rachis 7–8.5 cm. long, puberulous; flowers cream, nodding; bracts and bracteoles narrowly ovate, 1–1.5 mm. long, soon caducous; peduncle 5–25 mm.

long, sparsely puberulous; pedicels 1–3 mm. long, puberulous. Sepals ovate, ± 2.5 mm. long, obtuse, margin ciliate. Petals oblong-oval, 4.5–5 mm. long, glabrous, not recurved at the apex. Stamens 3 mm. long, the filaments fused together for all or most of their length into a tube with a free, rather irregular margin; connective-appendage oval, 1.5 times as long as the anther; thecal appendage short, bilobed. Ovary glabrous; style 3 mm. long. Capsule 3-lobed, 11–12 mm. long, coriaceous, smooth, glabrous. Fig. 3/10–12.

TANZANIA. Lushoto District: E. Usambara Mts., Ngwelo [Nguelo], *Scheffler* 26! & Sengale [Singale], Feb. 1932, *Greenway* 2910! & Muheza–Amani road, near Sigi, Mar. 1974, *Faden* 74/341!
DISTR. T 3; not known elsewhere
HAB. Lowland evergreen forest; 430–800 m.

NOTE. A little known species which Tennant (K.B. 16: 422 (1963)) places in *R. brachypetala* (Turcz.) Kuntze, however, the two are very different in a number of characters. *R. scheffleri* finds its closest ally in *R. subintegrifolia* and may perhaps be of hybrid origin between this and another species. *R. brachypetala* is only known from Bukoba District in Tanzania, whereas *R. subintegrifolia* is common in parts of E. and NE. Tanzania.

8. R. dentata *(P. Beauv.) Kuntze,* Rev. Gen. Pl. 1: 42 (1891); Hiern, Cat. Afr. Pl. Welw. 1: 36 (1896); Th. Dur. & Schinz, Consp. Fl. Afr. 1(2): 209 (1898); De Wild. & Th. Dur., Ann. Mus. Congo, Bot, sér 3(1): 10 (1901); De Wild. in B.J.B.B. 6: 148, 159 (1920); V.E. 3(2): 552 (1921); Melchior in E. & P. Pf., ed. 2, 21: 351, fig. 151G (1925); Exell & Medonça, C.F.A. 1: 74 (1937); T.T.C.L.: 646 (1949); I.T.U., ed. 2: 446 (1952), pro parte; F.W.T.A., ed. 2, 1: 101, 104 (1954); Tennant in K.B. 16: 425 (1963); Taton in F.C.B., Violaceae: 35 (1969); Grey-Wilson in K.B. 36: 116 (1981); Hamilton, Ug. For. Trees: 130 (1981). Type: Nigeria, near Buonopozo, Warri [Oware], *Palisot de Beauvois* (G, holo.)

Shrub or small tree to 6 m., though often less; young branches pubescent, but soon becoming glabrescent. Leaf-lamina oblong to elliptic or oblong-obovate, 8–24(–28) cm. long, 2.7–8.5(–10) cm. wide, the base cuneate to subrounded, the apex acuminate, glabrous above, often with some hairs at least on the midrib and lateral veins beneath, margin finely serrate-dentate, lateral veins 8–12 pairs; petiole 5–12 mm. long, pubescent to puberulous. Inflorescence a terminal thyrse, often broadly pyramidal or subcorymbose in general outline; rachis usually 5–12 cm. long, puberulous, rarely shortly pubescent; flowers yellow or greenish yellow; bracts and bracteoles narrowly ovate to linear-lanceolate, 1–4 mm. long, margin ciliate, soon caducous; peduncle 10–32 mm. long, usually puberulous; pedicels slender, 1–3 mm. long, puberulous. Sepals lanceolate to triangular-ovate, 2–2.5 mm. long, glabrous or puberulous, margin ciliate. Petals elliptical to elliptical-lanceolate, 3.5–4 mm. long, slightly recurved at the apex, often puberulous on the outside. Stamens 3.5–3.7 mm. long, the filaments fused together into a tube in the lower half, the tube with a partly free, lobed or uneven, margin; connective-appendage ovate, slightly shorter than the anthers; thecal appendages 2, narrow-lanceolate. Ovary scurfy; style 2.5–3 mm. long. Capsule 3-lobed, 10–18 mm. long, scurfy. Fig. 3/6–9.

UGANDA. Masaka District: Minziro Forest, *Wright Hill* 65!; Mengo District: Mabira Forest, Mar. 1900, *Ussher* 88! & Mulange, May 1919, *Dummer* 4106!
TANZANIA. Bukoba District: Kikuru Forest, Sept.–Oct. 1935, *Gillman* 445!
DISTR. U 4; T 1; Liberia, Ghana, Nigeria, Cameroun, Fernando Po, Zaire, Angola
HAB. Evergreen forest, sometimes in semi-swamp forest; 1200–1300 m.

SYN. *Ceranthera dentata* P. Beauv., Fl. Owar. 2: 11, t. 65 (1808): DC., Prodr. 1: 314 (1824)
    Alsodeia dentata (P. Beauv.) Oliv., F.T.A. 1: 110 (1868); Engl., P.O.A. A: 91 (1895); Th. & H. Dur., Syll. Fl. Congo: 34 (1909)

Rinorea bipindensis Engl. in E.J. 33: 145 (1902); Brandt in E.J. 50, Suppl.: 417 (1914); De Wild. in B.J.B.B. 6: 147, 152 (1920); V.E. 3(2): 552 (1921); Melchior in E.& P. Pf., ed. 2, 21: 351 (1925). Type: Cameroun, Bipindé, Epossi, *Zenker* 1599 (B, holo.†, K, iso.!)

NOTE. *R. dentata* is often confused with *R. ferruginea* in herbaria, however, the two are quite distinct and do not overlap in distribution. The former can be usually distinguished by its finely toothed leaves, the puberulous inflorescence and scurfy nature of the ovary and fruit; *R. ferruginea* has more coarsely toothed leaves, distinctly rusty-pubescent inflorescences and pubescent ovaries and fruits.

Several other names probably also belong here including *R. dewevrei* Engl., *R. brieyi* De Wild. and *R. kionzoensis* De Wild., though these require further investigation.

9. R. ferruginea *Engl.* in E.J. 33: 144 (1902); Brandt in E.J. 50, Suppl.: 417 (1914); De Wild. in B.J.B.B. 6: 147, 162 (1920); V.E. 3(2): 552 (1921); Melchior in E. & P. Pf., ed. 2, 21: 351 (1925); T.T.C.L.: 646 (1949); Tennant in K.B. 16: 426 (1963). Type: Tanzania, Uluguru Mts., near Tegetero, *Stuhlmann* 9036 (B, holo.†)

Shrub or small tree to 6 m. tall, though often less; young branches rusty-pubescent, sometimes glabrescent or subglabrous. Leaf-lamina oblanceolate to oblong or obovate, 8–25(–30) cm. long, 3–8.5 cm. wide, subrounded to cuneate at the base, abruptly to ± gradually acuminate at the apex, glabrous or with a few hairs on the midrib above, generally pubescent on the midrib and lateral veins beneath, often only sparsely so, margin serrate to serrate-dentate, subentire towards the base, lateral veins 12–17 pairs; petiole 0.8–5.3 cm. long, pubescent, rarely subglabrous. Inflorescence a terminal broad thyrse, pyramidal or subcorymbose generally, the rachis 4–10 cm. long; flowers yellowish, whitish yellow or greenish white, nodding in bud but usually becoming erect eventually; bracts and bracteoles ovate-triangular, 0.5–3 mm. long, pubescent, persistent; peduncle 10–30 mm. long, pubescent; pedicels 1–3 mm. long, pubescent. Sepals ovate, 2–3 mm. long, glabrous or pubescent along the midrib, ciliate. Petals ovate-lanceolate, 3.5–5 mm. long, often pubescent in the lower half outside, ciliate, not recurved at the apex. Stamens 2.5–3.5 mm. long, the filaments joined into a tube with a free irregular margin; connective-appendage oval, decurrent, up to twice the length of the anthers, subentire; thecal appendages 2, lanceolate. Ovary pubescent; style 2.5–3 mm. long. Capsule 3-lobed, 11–17 mm. long, shortly pubescent or glabrescent, rugose, blackish brown when ripe. Fig. 3/1–5.

KENYA. Kwale District: Shimba Hills, Kwale Forest, Mar. 1968, *Magogo & Glover* 405! & Mwele Mdogo Forest, Feb. 1953, *Drummond & Hemsley* 1104! & Aug. 1953, *Drummond & Hemsley* 3963!
TANZANIA. Lushoto District: Amani, track to Bomole, Apr. 1968, *Renvoize* 1618!; Morogoro District: Nguru Mts., Manyangu Forest, N. of Liwale R., Apr. 1953, *Drummond & Hemsley* 1970! & Uluguru Mts., Kitundu, Nov. 1934, *E.M. Bruce* 194!; Zanzibar I., Chwaka, Aug. 1963, *Faulkner* 3266!
DISTR. K 7; T 3, 6–8; Z; Mozambique, Zambia and Zimbabwe
HAB. Evergreen forest; 150–1300 m.

SYN. *R. ferruginea* Engl. var. *heinsenii* Engl. in E.J. 33: 144 (1902); T.T.C.L.: 647 (1949). Type: Tanzania, E. Usambara Mts., Derema [Nderema], *Heinsen* 37 (B, holo.†)
Alsodeia gazensis Bak.f. in J.L.S. 40: 22 (1911). Lectotype, chosen here: Zimbabwe, Chirinda Forest, *Swynnerton* 132 (K, lecto.!, BM, isolecto.!)
Rinorea zimmermannii Engl. in E.J. 51: 121 (1913); Brandt in E.J. 50, Suppl.: 416 (1914); De Wild. in B.J.B.B. 6: 147, 193 (1920); V.E. 3(2): 552 (1921); Melchior in E. & P. Pf., ed. 2, 21: 351 (1925). Types: Tanzania, E. Usambara Mts., Amani, *Zimmermann in Herb. Amani* 1460 & 1576 & *Braun in Herb. Amani* 1665 (B, syn. †, EA, isosyn.)
R. usambarensis Engl. in E.J. 51: 126 (1913); Brandt in E.J. 50, Suppl.: 417 (1914); De Wild. in B.J.B.B. 6: 149, 189 (1920); V.E. 3(2): 553 (1921); Melchior in E. & P. Pf.,

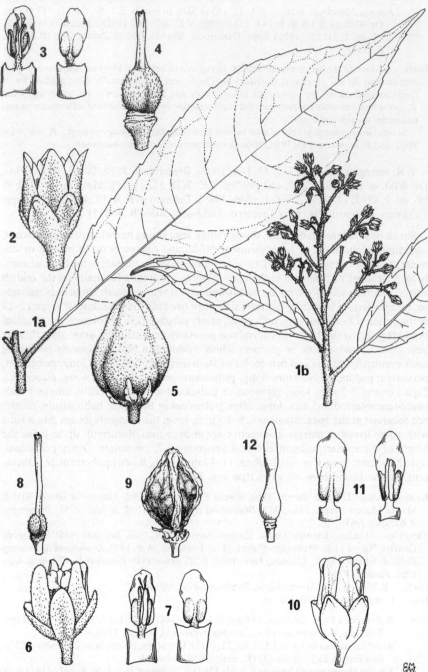

FIG. 3. *RINOREA FERRUGINEA* – **1,** habit, x $\frac{2}{3}$; **2,** flower, x 4; **3,** stamens, anterior and posterior views, x 6; **4,** ovary and style, x 6; **5,** fruit x 2. *R. DENTATA* – **6,** flower, x 4; **7,** stamens, anterior and posterior views, x 6; **8,** ovary and style, x 6; **9,** fruit, x 2. *R. SCHEF-FLERI* – **10,** flower, x 4; **11,** stamens, anterior and posterior views, x 6; **12,** ovary, x 6. 1–4, from *Drummond & Hemsley* 1970; 5, from *Drummond & Hemsley* 3022; 6–8, from *Brasnett* 65; 9, from *Devred* 403; 10–12, from *Greenway* 2910. Drawn by Christine Grey-Wilson.

ed. 2, 21: 351 (1925); T.T.C.L.: 647 (1949). Types: Tanzania, E. Usambara Mts., Amani, *Braun* in *Herb. Amani* 1985, *Engler* 3405, *Warnecke* 256 (B, syn.†, EA, isosyn.), *Zimmermann* in *Herb. Amani* 1147, *Engler* 738 & *Holtz* 740 (B, syn.†)

R. gazensis (Bak.f.) Brandt in E.J. 50, Suppl.: 416 (1914); De Wild. in B.J.B.B. 6: 147, 162 (1920); V.E. 3(2): 552 (1921); Robson in F.Z. 1: 253, t. 39C/1–3 (1960); K.T.S.: 599 (1961)

NOTE.   *R. ferruginea* hybridises with *R. angustifolia* subsp. *albersii, R. elliptica* and *R. squamosa*, see p. 23.

10. R. tshingandaensis *Taton* in B.J.B.B. 38: 384 (1968) & in F.C.B., Violaceae: 33 (1969); Grey-Wilson in K.B. 36: 121 (1981). Type: Zaire, Kavumu–Walikale, Tshinganda R., *Pierlot* 2463 (BR, holo.)

Shrub or small tree to 15 m., though generally less; young branches glabrous or sparsely pubescent. Leaf-lamina elliptic to elliptic-oblanceolate, 4.8–14.5 cm. long. 1.6–4.8 cm. wide, the base ± cuneate, the apex acuminate, glabrous, margin shallowly serrate-crenate to subentire, lateral veins 6–9 pairs; petiole 6–14 mm. long, glabrous. Inflorescence a broad terminal thyrse, generally with fewer than 20 flowers, the rachis short, usually 15–32 mm. long, puberulous; flowers yellow or brownish yellow, nodding at maturity; bracts and bracteoles ovate to linear-lanceolate, 1–2 mm. long, margin ciliate; peduncle 5–11 mm. long, sparsely puberulous; pedicels slender, 2–3 mm. long, sparsely puberulous. Sepals ovate, 2–3 mm. long, margin ciliate. Petals oval-lanceolate to lanceolate, 5–6 mm. long, margin ciliate. Stamens 4–4.5 mm. long, the filaments fused together into a tube in the lower half, with a free rather irregular margin; connective-appendage elliptic-lanceolate, ± equal in length to the anthers; thecal appendages 2, long, ± two-thirds the length of the connective-appendage. Ovary glabrous; style 3–4 mm. long. Capsule 3-lobed, 14–18 mm. long, finely rugose. Fig. 4/1–4.

UGANDA.   Kigezi District: Kigezi, Apr. 1948, *Purseglove* 2671! & Ishasha Gorge, Aug. 1949, *Purseglove* 3057!
DISTR.   U 2; E. Zaire
HAB.   Upland evergreen forest; 1350–1500 m.

SYN.   [*R. dentata* sensu I.T.U., ed. 2: 446 (1952), pro parte, *non* (P. Beauv.) Kuntze]

11. R. convallarioides *(Bak.f.) Eyles* in Trans. Roy. Soc. S. Afr. 5: 421 (1916); Melchior in E. & P. Pf., ed. 2, 21: 350 (1925); N. Robson in F.Z. 1: 247 (1960); Grey-Wilson in K.B. 36: 114 (1981). Lectotype, chosen here: Zimbabwe, Chirinda Forest, *Swynnerton* 2119 (K, lecto., BM, isolecto.)

Shrub or small tree to 15 m., though generally far less; young branches finely pubescent or glabrous. Leaf-lamina elliptic to elliptic-oblong or elliptic-oblanceolate, 2–16 cm. long, 1.2–5.8 cm. wide, the base usually subrounded, the apex subobtuse to acuminate, glabrous or rarely pubescent on the midrib and lateral veins beneath, margin serrate to serrate-dentate or subentire, lateral veins 6–8 pairs; petiole 2–7 mm. long, bristly pubescent, sometimes glabrescent. Inflorescence an axillary raceme, usually solitary but sometimes several together at a node, or branched near the base; flowers ± erect, cream or greenish white, sometimes pink tinged; peduncle slender, 1–7.4 cm. long, finely pubescent; bracts ovate, cucullate, 1–2 mm. long, ribbed, persistent; pedicels 1.5–8 mm. long, pubescent. Sepals ovate, 1–1.5 mm. long, ribbed, ciliate. Petals oblong to ovate-lanceolate, 4–6.5 mm. long, somewhat reflexed at the tip, glabrous. Stamens 3–3.5 mm. long, free or fused together at the very base of the filaments; connective-appendage suborbicular-ovate, almost as long as the anthers;

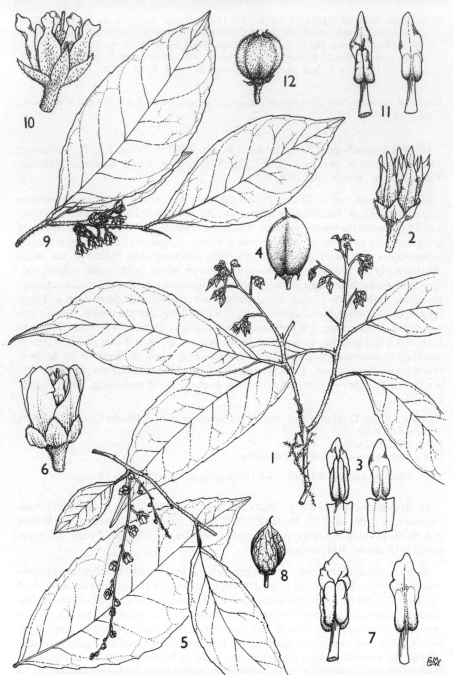

FIG. 4. *RINOREA TSHINGANDAENSIS* – **1,** habit, x ⅔; **2,** flower, x 4; **3,** stamens, anterior annd posterior views, x 6; **4,** fruit, x 1. *R. CONVALLARIOIDES* subsp. *MARSABITENSIS* – **5,** habit, x ⅔; **6,** flower, x 4; **7,** stamens, anterior and posterior views, x 8; **8,** fruit, x 1. *R. SQUAMOSA* subsp. *KAESSNERI* – **9,** habit, x ⅔; **10,** flower, x 4; **11,** stamens, anterior and posterior views, x 8; **12,** fruit, x 1. 1–3, from *Purseglove* 3057; 4, from *Purseglove* 2671; 5–7, from *Adamson* 9; 8, from *Bally & Smith* 14848; 9–11, from *Kassner* 310; 12, from *Semsei* 1482. Drawn by Christine Grey-Wilson.

thecal appendages 2, or 1 bifid. Ovary glabrous; style $\pm$ 3 mm. long. Capsule 3-lobed, 11–14 mm. long, coriaceous, glabrous, yellowish green when ripe.

SYN. *Alsodeia convallarioides* Bak.f. in J.L.S. 40: 21 (1911)

subsp. **marsabitensis** *Grey-Wilson* in K.B. 36: 115 (1981). Type: Kenya, Northern Frontier Province, Marsabit Forest, *T. Adamson* 9 (K, holo.!)

Leaf-lamina 2–9(–10) cm. long, 1.2–4.5 cm. wide, the apex subobtuse or very shortly acuminate; petiole 2–5 mm. long. Peduncle (including rachis) 3–7.4 cm. long. Fig. 4/5–8

KENYA. Northern Frontier Province: Marsabit Forest, Feb. 1953, *Gillett* 15101! & Jan. 1958, *T. Adamson* 9! & Sokota Dika to Gof Sokota Guda, Jan. 1972, *Bally & Smith* 14848!
DISTR. K 1; not known elsewhere
HAB. Montane evergreen forest; $\pm$ 1450–1500 m.

SYN. [*R. convallariiflora* sensu K.T.S.: 599 (1961), *non* Brandt]

NOTE. Subsp. *convallarioides* is native to Zimbabwe, Malawi and Mozambique. It has small leaves similar to subsp. *marsabitensis* but can be readily distinguished by its shorter, more densely crowded racemes, 1–2.6 cm. long. Subsp. *occidentalis* Grey-Wilson (*R. convallariiflora* Brandt) from W. Africa is an altogether more robust plant with larger, strongly acuminate leaves with petioles 5–7 mm. long.

12. R. beniensis *Engl.* in Z.A.E. 2: 561 (1913) & in E.J. 51: 107 (1913); Brandt in E.J. 50, Suppl.: 410 (1914); De Wild. in B.J.B.B. 6: 140, 152 (1920); V.E. 3(2): 547 (1921); Melchior in E. & P. Pf., ed. 2, 21: 350 (1925); Tennant in K.B. 16: 411 (1963); Taton in F.C.B., Violaceae: 13, fig. 2 (1969). Types: Zaire, Beni, Mwera [Muera], *Mildbraed* 2406 & 2768 & Mokoko, *Mildbraed* 2935 (B, syn.†)

Shrub or small tree to 12 m. tall; branches sparsely pubescent at first. Leaf-lamina elliptic to elliptic-lanceolate, 3.6–11.4 cm. long, 1.6–4.2 cm. wide, the base cuneate to subrounded, the apex acuminate to caudate, glabrous or with a few hairs on the midrib and lateral veins beneath at first, margin crenate-serrate to $\pm$ entire, lateral veins 6–9 pairs; petiole 3–4 mm. long, pubescent. Inflorescence a few-flowered axillary raceme, or solitary or fascicled; flowers white, fragrant; peduncle very uneven, slender, 2–30 mm. long, finely pubescent; bracts ovate, 1–2 mm. long, falling eventually; pedicels very slender, 4–19 mm. long, finely pubescent. Sepals oval, 1.5–2 mm. long, ribbed, ciliate. Petals narrow-oblong, 6.5–8 mm. long, ciliate at the top. Stamens 2.8–3.25 mm. long, free or fused together at the very base of the filaments; connective-apppendage suborbicular, as long as, but twice as broad as, the anther, margin irregularly dentate; thecal appendages absent. Ovary glabrous; style 3–3.5 mm. long. Capsule shallowly 3-lobed, 9–12 mm. long, glabrous, greenish yellow when ripe. Fig. 5/1–5.

UGANDA. Bunyoro District: Budongo Forest, Nov. 1935, *Eggeling* 2269! & Nov. 1938, *Loveridge* 109!; Toro District: Semliki Forest, near Kirimia R., Oct. 1951, *Osmaston* 1366! & 1367!
DISTR. U 2; E. Zaire
HAB. Rain-forest; 750–1150 m.

SYN. *Rinorea affinis* Robyns & Lawalrée in B.J.B.B. 18: 280 (1947); F.P.N.A. 1: 627 (1948). Type: Zaire, Semliki, *Bequaert* 3236 (BR, holo.!, K, iso.!)
[*R. ardisiiflora* sensu I.T.U., ed. 2: 446, fig. 94 (1952), as *'ardisiaeflora';* Hamilton, Ug. For. Trees: 130 (1981), *non* (Oliv.) Kuntze]

NOTE. *R. sp. A*, p.24, is possibly closely related to *R. beniensis*.

13. R. squamosa *(Tul.) Baill.* in Bull. Soc. Linn. Paris 1: 583 (1886); Perrier in Fl. Madag., Fam. 139: 34 (1955). Types: Madagascar, Sambirano, Nosy-Be, *Boivin* 2122 (P, syn.) & *Pervillé* 255 (P, syn., K, isosyn.!)

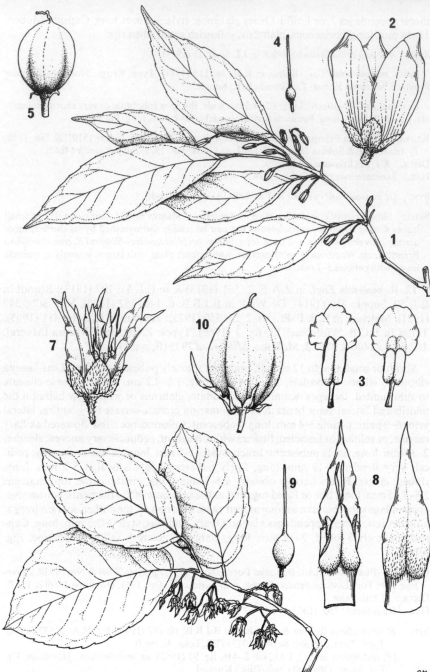

Fig. 5.   *RINOREA BENIENSIS* – **1**, habit, x $\frac{2}{3}$; **2**, flower, x 4; **3**, stamens, anterior and posterior
views, x 8; **4**, ovary and style, x 4; **5**, fruit, x 3. *R. ELLIPTICA* – **6**, habit, x $\frac{2}{3}$; **7**, flower,
x 4; **8**, stamens, anterior and posterior views, x 8; **9**, ovary and style, x 4; **10**, fruit, x 3. 1–4,
from *Eggeling* 2267; 5, from *Osmaston* 1367; 6–9, from *Semsei* 3134; 10, from *Bally* 2043.
Drawn by Christine Grey-Wilson.

Shrub or small tree to 4 m. tall; branches glabrous or sparsely pubescent at first. Leaf-lamina elliptic to elliptic-oblong or lanceolate-elliptic, 2.5–14.5 cm. long, 1–6.2 cm. wide, the base rounded or subrounded to ± cuneate, the apex subacute to acuminate, glabrous or pubescent on the midrib beneath, margin usually finely serrate-dentate, lateral viens 6–10 pairs; petiole 3–8 mm. long, pubescent. Inflorescence a short axillary raceme, often appearing subcorymbose or fasciculate; flowers greenish white, sometimes flushed with pink inside; peduncle 3–10(–15) mm. long, pubescent; bracts ovate, 2–3 mm. long, closely imbricate at first, pubescent, persistent; pedicels slender, 3–6 mm. long, pubescent. Sepals ovate, 1.5–2.5 mm. long, finely ribbed, pubescent, persistent. Petals oblong-oblanceolate, 3.5–5 mm. long, ciliate, often pubescent on the exterior, the tip somewhat recurved. Stamens 3–3.2 mm. long, usually fused together at the very base of the filaments; connective-appendage oblong, entire, slightly shorter and scarcely wider than the anther. Ovary glabrous to densely pubescent; style 2–2.5 mm. long. Capsule ± globose, scarcely lobed, rugose, glabrous or pubescent.

SYN.   *Alsodeia squamosa* Tul. in Ann. Sci. nat., sér. 5, 9: 307 (1868)

subsp. **kaessneri** *(Engl.) Grey-Wilson* in K.B. 36: 120 (1981). Type: Kenya, Kwale District, Bome R., *Kässner* 310 (B, holo.†, K, iso.!)

Leaf-lamina 4.7–14.5 cm. long, 2.6–6.2 cm. wide, lateral veins 6–10 pairs; petiole 3–8 mm. long. Ovary and fruit-capsule pubescent. Fig. 4/9–12.

KENYA.   Kwale District: Bome R., Mar. 1902, *Kassner* 310! & between Umba and Mwena Rivers, Aug. 1953, *Drummond & Hemsley* 3858!; Lamu District: Utwani Forest, Dec. 1934, *Mohamed Abdullah* in *F.D.* 3346!
TANZANIA.   Morogoro District: Nguru Mts., Ruhamba [Koruhamba], Nov. 1953, *Paulo* 210! & Uluguru Mts., Turiani, Sept. 1953, *Eggeling* 6708! & Nov. 1953, *Semsei* 1482!
DISTR.   K 7; T 6; not known elsewhere
HAB.   Evergreen lowland and submontane forest; 30–450 m.

SYN.   *R. kaessneri* Engl. in E.J. 34: 317 (1904); K.T.S.: 601 (1961)

NOTE.   The typical subspecies is endemic to Madagascar, differing from subsp. *kaessneri* by its rather smaller leaves and glabrous ovaries and fruit-capsules. This species hybridises with *R. ferruginea* in Tanzania, p. 24.

14. **R. elliptica** *(Oliv.) Kuntze,* Rev. Gen. Pl. 1: 42 (1891); Reiche & Taub. in E. & P. Pf. III. 6: 329 (1895); Th. Dur. & Schinz, Consp. Fl. Afr. 1(2): 210 (1898); Brandt in E.J. 50, Suppl.: 411 (1914); De Wild. in B.J.B.B. 6: 140, 161 (1920); V.E. 3(2): 548 (1921); Melchior in E. & P. Pf., ed. 2, 21: 350, fig. 151D (1925); T.T.C.L.: 645 (1949); N. Robson in F.Z. 1: 248 (1960); K.T.S.: 599 (1961); Tennant in K.B. 16: 417 (1963). Type: Mozambique/Tanzania, Ruvuma [Rovuma] R., 32 km. from mouth, *Kirk* (K, holo.!)

Shrub or small tree to 12 m. tall; branches slender, glabrous. Leaf-lamina broadly elliptic to elliptic-oblong or elliptic-lanceolate, 3.4–12.7 cm. long, 2–6.7 cm. wide, the base subcordate to rounded, the apex usually subobtuse or subacute, sometimes shortly acuminate, glabrous, margin crenate-serrate to serrulate, lateral veins 5–11 pairs, prominent; petiole 7–10 mm. long, glabrous or slightly pubescent. Inflorescence a rather lax axillary raceme, sometimes branched at the base, with 2–10 flowers generally; flowers erect or spreading, white or pale pink, aromatic; peduncle 11–37 mm. long, sparsely pubescent, puberulous or glabrous; bracts and bracteoles oblong-elliptic, 1.5–2.5 mm. long, ribbed, margin ciliate, soon caducous; pedicels slender, 4–14 mm. long, sparsely pubescent to glabrous. Sepals ovate to ovate-lanceolate, 1.5–3 mm. long, ribbed, glabrous or sparsely pubescent, margin ciliate. Petals elliptic-lanceolate, 6–10 mm. long, erect or spreading but only tardily reflexing, margin ciliate. Stamens 5–6 mm. long, the filaments fused together completely to form a tube; connective-appendage lanceolate, entire, decurrent, almost twice as long as the anther; thecal appendages 2, linear-lanceolate. Ovary glabrous. Capsule subglobose, 6–9 mm. long, obscurely 3-lobed, glabrous, smooth, reddish when ripe. Fig. 5/6–10.

KENYA. Tana River District: Bura, Jan. 1943, *Bally* 2043!; Lamu District: Utwani Forest, Sept. 1943, *Mohamed Abdullah* in *F.D.* 3349! & Utwani, Ndogo Forest, Aug. 1956, *Rawlins* 70!
TANZANIA. Lushoto District: Mombo, Nov. 1955, *Milne-Redhead & Taylor* 7268!; Pangani District: 5 km. W. of Magunga Halt, Pangani R., Dec. 1960, *Semsei* 3134!; Morogoro District: Lusunguru Forest Reserve, 6.5 km. NE. of Turiani, Mar. 1953, *Drummond & Hemsley* 1926!
DISTR. K 7; T 2, 3, 6, 8; Mozambique, Malawi
HAB. Lowland evergreen forest; 50–600 m.

SYN. *Alsodeia elliptica* Oliv., F.T.A. 1: 108 (1868)
    A. usambarensis Engl. in Abh. Preuss. Akad. Wiss.: 36 (1894), *nomen nudum*

NOTE. *Rinorea comorensis* Engl. (*R. hildebrandtiana* H. Perrier, *nom. illegit.*) was considered by Tennant, K.B. 16: 417 (1963), to be conspecific with *R. elliptica*. However, the Comoro Islands plant has relatively longer and narrower leaves which are acuminate, and distinctly longer inflorescences (5–10 cm. long). I have assigned this plant to *R. elliptica* subsp. *comorensis* (Engl.) Grey-Wilson. *R. elliptica* hybridises with *R. ferruginea*, p. 23.

15. R. angustifolia *(Thouars) Baill.* in Bull. Soc. Linn. Paris 1: 582 (1886); Kuntze, Rev. Gen. Pl. 1: 42 (1891); Th. Dur. & Schinz, Consp. Fl. Afr. 1(2): 208 (1898); Melchior in E. & P. Pf., ed. 2, 21: 352 (1925); H. Perrier in Mém. Inst. Sci. Madag., sér. B, 2: 328 (1949) & in Fl. Madag., Fam. 139: 36 (1955); Tennant in K.B. 16: 412 (1963); Grey-Wilson in K.B. 36: 110 (1981). Type: Madagascar, no precise locality, *Du Petit Thouars* (P, holo.)

Shrub or small tree to 10 m. tall; stems glabrous or pubescent at first. Leaf-lamina narrowly elliptic to oblong-elliptic, oblong or oblong-ovate, 2.3–13 cm. long, 1–4.8 cm. wide, the base ± cuneate, subrounded to subcordate, the apex obtuse to acuminate, glabrous or with a few hairs on the midrib above and/or beneath, margin serrate-crenate to subentire, lateral veins 5–11 pairs; petiole 1–9 mm. long, glabrous or pubescent. Inflorescence a simple axillary raceme carrying 1–20 flowers usually, occasionally 2 racemes per node; flowers white or yellowish white, fragrant, erect to nodding; peduncles 2–7 cm. long, brownish, densely pubescent; bracts ovate to ovate-oblong, 0.5–2 mm. long, ribbed, pubescent, persistent; pedicels slender, 4–15 mm. long, pubescent. Sepals ovate to oblong, 1.5–2.5 mm. long, ribbed, pubescent in the middle, margin usually ciliate. Petals lanceolate, 3–6.5 mm. long, acute, strongly reflexed soon after anthesis, glabrous or pubescent in the centre on the exterior, sometimes ciliate towards the top. Stamens 2.5–6 mm. long, the short filaments completely fused into a ring with a free ciliate margin; connective-appendage lanceolate, acute, decurrent, ± twice as long as the anther; thecal appendages 2, sometimes solitary, linear-lanceolate, short. Ovary usually pubescent; style 2.5–4 mm. long. Capsule subglobose, 6–9 mm. long, poorly 3-lobed, sparsely to moderately pubescent or glabrous.

SYN. *Alsodeia angustifolia* Thouars, Hist. Vég. Isles Austr. Afr.: 57, t. 18/1 (1805)

NOTE. A polymorphic species widespread in west, central and east Africa and Madagascar. It can be divided into a number of subspecies on characters of the leaves and inflorescence. However, the exact distribution and variation of the taxa involved needs more critical investigation.

Petiole glabrous, rarely slightly pubescent (**T** 3) .   .   .   .   .   . b. subsp. **albersii**

Petiole pubescent:
  Leaf-lamina usually long-acuminate, with 8–12 pairs of lateral
    veins; petiole 3–8 mm. long (U 2, T 1)   .   .   .   .   . c. subsp. **engler-**
                                                                       iana
  Leaf-lamina subacute to shortly acuminate, with 6–8 pairs of
    lateral veins; petiole 1–4 mm. long (K 7, T 6–8)   .   .   . a. subsp. **ardisii-**
                                                                       flora

a. subsp. **ardisiiflora** *(Oliv.) Grey-Wilson* in K.B. 36: 113 (1981). Lectotype (see C.F.A.): Angola, Cuanza Norte, Pungo Andongo, Mata de Pungo, *Welwitsch* 885 (LISU, lecto., BM, K, isolecto.!)

Young stems pubescent, sometimes subglabrous. Leaf-lamina 2.3–8.2 cm. long, 1–2.9 cm. wide, lateral veins 6–8 pairs; petiole 1–4 mm. long, pubescent. Mature inflorescence 1.5–4 cm. long. Fig. 6/7, 8.

KENYA.   Kilifi District: Ribe Kaya, Apr. 1981, *Hawthorne* 327!
TANZANIA.   Uzaramo District: Pugu Hills, Mar. 1964, *Semsei* 3677!; Iringa District: W. Mufindi, Nov. 1947, *Brenan & Greenway* 8257!; Lindi District: Lake Lutamba, Dec. 1937, *Schlieben* 5713!
DISTR.   K 7; T 6–8; Angola, Mozambique, Malawi, Zambia
HAB.   Evergreen forest; 100–2200 m.

SYN.   *Alsodeia ardisiiflora* Oliv., F.T.A. 1: 108 (1868), as *'ardisiaeflora'*
       Rinorea ardisiiflora (Oliv.) Kuntze, Rev. Gen. Pl. 1: 42 (1891); Reiche & Taub. in E.
           & P. Pf. III. 6: 329 (1895); Th. Dur. & Schinz, Consp. Fl. Afr. 1(2): 208 (1898); Brandt
           in E.J. 50, Suppl.: 411 (1914); De Wild. in B.J.B.B. 6: 141, 150 (1920); Melchior in
           E. & P. Pf., ed. 2, 21: 350 (1925); Exell & Mendonça, C.F.A. 1: 74 (1937)
       R. holtzii Engl., E.J. 34: 317 (1904); Brandt in E.J. 50, Suppl.: 411 (1914); De Wild.
           in B.J.B.B. 6: 140, 166 (1920); V.E. 3(2): 548 (1921); T.T.C.L.: 645 (1949); N. Robson
           in F.Z. 1: 248, t. 39A/1–6 (1960), pro parte. Type: Tanzania, Uzaramo District, Pugu
           Hills, *Holtz* 660 (B, holo.†, EA, iso.)
       R. sp. A sensu N. Robson in F.Z. 1: 254 (1960)

NOTE.   *R. myrsinifolia* Dunkley from Malawi is a plant with unusually small leaves and single-flowered inflorescences which I have assigned to subsp. *myrsinifolia* (Dunkley) Grey-Wilson in K.B. 36: 114 (1981). The South African *R. natalensis* Engl. likewise is geographically distinct, differing from all its allies in having elliptic-rhombic, sinuate-dentate margined leaves and from the E. African plants in having glabrous ovaries and fruits. However, there are few other differences and this taxa has been assigned subspecific rank – subsp. *natalensis* (Engl.) Grey-Wilson. The Madagascan *R. angustifolia* (Thouars) Baill. sensu stricto, has rather small leaves, not unlike subsp. *ardisiiflora*, but it can be readily distinguished by its smaller flowers and long-acuminate leaves. *Mahberley* 1272 from the Ukaguru Mts., Mamiwa Forest Reserve, probably belongs to subsp. *ardisiiflora*, but the material is rather immature and needs further collecting.

b. subsp. **albersii** *(Engl.) Grey-Wilson* in K.B. 36: 113 (1981). Type: Tanzania, Lushoto District, W. Usambara Mts., Kwai, *Albers* 315 (B, holo.†)

Young stems glabrous. Leaf-lamina 4.5–11.5 cm. long, 2–4.8 cm. broad, the apex subacute to shortly acuminate, lateral veins 5–9 pairs; petiole 5–9 mm. long, glabrous, rarely slightly pubescent. Mature inflorescence 3.5–7 cm. long. Fig. 6/3–6.

TANZANIA.   Lushoto District: Usambara Mts., Kwamkoro–Sangarawe, Aug. 1929, *Greenway* 1725! & 6.5 km. NE. of Lushoto, Mkuzi, Apr. 1953, *Drummond & Hemsley* 2065! & W. Usambara Forests, *D.K. Grant* 27!
DISTR.   T 3; not known elsewhere
HAB.   Evergreen forest; 950–1750 m.

FIG. 6. *RINOREA ANGUSTIFOLIA* subsp. *ENGLERIANA* – **1,** habit, x $\frac{2}{3}$; **2,** fruit, x 2. *R. ANGUSTIFOLIA* subsp. *ALBERSII* – **3,** habit, x $\frac{2}{3}$; **4,** flower, x 4; **5,** stamens, anterior and posterior views, x 6; **6,** fruit, x 2. *R. ANGUSTIFOLIA* subsp. *ARDISIIFLORA* – **7,** habit, x $\frac{2}{3}$; **8,** leaf, upper surface, x $\frac{2}{3}$. 1, 2, from *Gillman* 101; 3–5, from *Drummond & Hemsley* 2065; 6, from *Greenway* 7928; 7, from *Brenan & Greenway* 8257; 8, from *Semsei* 3677. Drawn by Christine Grey-Wilson.

SYN.   *R. albersii* Engl. in E.J. 33: 135 (1902); Brandt in E.J. 50, Suppl.: 412 (1914); De Wild.
        in B.J.B.B. 6: 141, 150 (1920); V.E. 3(2): 548 (1921); T.T.C.L.: 645 (1949)

NOTE.   Subsp. *albersii* hybridises with *R. ferruginea* (see below).

c. subsp. **engleriana** *(De Wild. & Th. Dur.) Grey-Wilson* in K.B. 36: 111 (1981). Type: Zaire,
Rewa, *Dewèvre* 1141 (BR, holo.)

Young stems glabrous, or pubescent but then soon glabrescent. Leaf-lamina 6–15 cm. long,
2.2–5.8 cm. wide, the apex usually long-acuminate, lateral veins 8–12 pairs; petiole 3–8 mm.
long, pubescent. Mature inflorescence usually 5.5–7 cm. long. Fig. 6/1, 2.

UGANDA.   Ankole District: Kalinzu Forest, Jan. 1953, *Osmaston* 2783!
TANZANIA.   Bukoba District: Kiau I., Aug. 1934, *Gillman* 101!
DISTR.   U 2; T 1; W. Tropical Africa, Cameroun, Zaire
HAB.   Evergreen forest; 1150–1500 m.

SYN.   *Alsodeia engleriana* De Wild. & Th. Dur. in B.S.B.B. 38: 172 (1900) & Mat. Fl. Congo
        6: 2 (1900)
        Rinorea engleriana (De Wild. & Th. Dur.) De Wild. & Th. Dur. in Ann. Mus. Congo,
        Bot., sér. 3(1): 11 (1901); Brandt in E.J. 50, Suppl.: 412 (1914); De Wild. in B.J.B.B.
        6: 141, 161 (1920); V.E. 3(2): 548 (1921)
        R. gracilipes Engl. in E.J. 33: 136 (1902); Brandt in E.J. 50, Suppl.: 412 (1914); De Wild.
        in B.J.B.B. 6: 141, 166 (1920); V.E. 3(2): 548 (1921); Melchior in E. & P. Pf., ed. 2,
        21: 349, fig. 150, 350 (1925); Exell & Mendonça, C.F.A. 1: 75 (1937); F.W.T.A., ed.
        2, 1: 99, 101 (1954); Taton in F.C.B., Violaceae: 17 (1969). Type: Cameroun, Bipindé,
        Zenker 1244 (K, lecto.!, B, BM, isolecto.)
        R. aruwimensis Engl. in Z.A.E. 2: 562 (1913) & in E.J. 51: 109 (1913); Brandt in E.J.
        50, Suppl.: 412 (1914); De Wild. in B.J.B.B. 6: 141, 151 (1920); V.E. 3(2): 548 (1921).
        Types: Zaire, Aruwimi, Jambuja, *Mildbraed* 3291 & 3291a (B, syn.†)
        [*R. ardisiiflora* sensu F.W.T.A., ed. 2, 1: 99, 101 (1954) & Taton in F.C.B., Violaceae:
        16 (1969), *non* (Oliv.) Kuntze sensu stricto]

Hybrids

The following 3 hybrids need authentication by detailed field work. The parentage
is a suggestion, although it seems almost certain that *R. ferruginea* is involved in each.

R. angustifolia *(Thouars) Baill.* subsp. **albersii** *(Engl.) Grey-Wilson* x **R. ferruginea**
Engl.; Grey-Wilson in K.B. 36: 122 (1981)

Leaf-lamina lyrate-oblanceolate, 6.2–15 x 1.8–5 cm., the base ± rounded, the apex
acuminate, pubescent in the angles between the midrib and lateral veins beneath,
lateral veins 9–11 pairs; petiole 4–7 mm. long, finely pubescent. Inflorescence a terminal or subterminal raceme, or axillary; bracts apparently caducous. Sepals 1.5 mm.
long, pubescent in the centre outside, margin ciliate. Petals 6.5–7 mm. long, strongly
reflexed at anthesis. Stamens 6–7 mm. long, the connective-appendage ± lanceolate,
twice as long as and broader than the anther. Ovary pubescent. Capsule unknown.

TANZANIA.   Lushoto District: E. Usambara Mts., Monga, *Zimmermann* in Herb. Amani 2513!
DISTR.   T 3; not known elsewhere

SYN.   *R. subumbellata* Brandt in E.J. 51: 110 (1913); T.T.C.L.: 645 (1949). Type: Tanzania,
        E. Usambara Mts., Monga, *Zimmermann* in Herb. Amani 2513 (B, holo.†, HBG, K,
        iso.!)

R. elliptica *(Oliv.) Kuntze* x **R. ferruginea** *Engl.*; Grey-Wilson in K.B. 36: 122 (1981)

Leaf-lamina lyrate-lanceolate, 4.6–12.8 x 1.7–4.8 cm., the base subcordate, the apex acuminate, pubescent mainly along the midrib beneath, lateral veins 8–10 pairs; petiole 1–2 mm. long, pubescent. Inflorescence an axillary or subterminal raceme; bracts persistent, pubescent. Sepal ± 1.25 mm. long, margin ciliate. Petals ± 5.5 mm. long, scarcely reflexed at anthesis. Stamens ± 4 mm. long, the connective-appendage lanceolate, equal in length to the anther. Ovary glabrous. Capsule 12–13 mm. long, 3-lobed, glabrous, smooth.

TANZANIA.   Morogoro District: Nguru Mts., Turiani, Nov. 1955, *Semsei* 1480!
DISTR.   T 6; not known elsewhere

R. ferruginea *Engl.* x **R. squamosa** *(Tul.) Baill.* subsp. **kaessneri** *(Engl.) Grey-Wilson;* Grey-Wilson in K.B. 36: 123 (1981)

Leaf-lamina oblong-oblanceolate, 5.5–13 x 1.8–4.6 cm., the base ± rounded, the apex acuminate, glabrous except for tufts of hairs in the angles formed between the midrib and lateral veins beneath, lateral veins 7–9 pairs; petiole 3–6 mm. long, pubescent. Inflorescence apparently an axillary raceme; bracts caducous. Flowers unknown. Ovary and young fruit densely bristly pubescent.

TANZANIA.   Morogoro District: Nguru Mts., Mtibwa, Nov. 1953, *Paulo* 202!
DISTR.   T 6; not known elsewhere

Imperfectly known species

R. sp. A

Tall tree, height unknown; stems slender, glabrous. Leaf-lamina elliptic to elliptic-oblong, 4.3–12.5 cm. long, 1.8–5 cm. wide, the base shortly cuneate, the apex acuminate, glabrous above but with a few short hairs on the midrib beneath, margin shallowly serrate-crenate to subentire, lateral veins 7–10 pairs; petiole 3–5 mm. long, pubescent. Inflorescence a very short axillary raceme, the rachis 2–4 mm. long; bracts lanceolate, ± 1.5 mm. long, ribbed, persistent; pedicels 15–18 mm. long, slender, sparsely pubescent. Flowers unknown. Capsule shallowly 3-lobed, 10–11 mm. long, glabrous, heavily veined.

TANZANIA.   Morogoro District: Uluguru Mts., Kimboza Forest Reserve, July 1952, *Semsei* 746!
DISTR.   T 6; not known elsewhere

NOTE.   This species comes closest to *R. beniensis* Engl. and indeed it may prove to be a subspecies of it. Flowering material is needed for comparison.

R. sp. B

UGANDA.   Masaka District: Kirala Forest, 1200 m., Jan. 1955, *Bunzinya* 2!; Lake Basin Forests, Nov. 1913, *Fyffe* 73!

NOTE.   The two specimens cited above appear to match *R. ituriensis* Brandt in vegetative characters, however the former has only immature flowers and no fruits and the latter is sterile. *R. ituriensis* is native to central Zaire.

Excluded species

R. cafassi *Chiov.,* Racc. Bot. Miss. Consol., Kenya: 6 (1935) = *Casearia battiscombei* R.E. Fries (Flacourtiaceae)

2. VIOLA

L., Sp. Pl.: 933 (1753) & Gen. Pl., ed. 5: 402 (1754)

Herbs or rarely subshrubs. Leaves alternate, peiolate; stipules entire to variously toothed or lobed, sometimes foliaceous. Flowers solitary, axillary, zygomorphic; pedicels with a pair of small bracteoles, generally above the middle. Sepals ± equal, usually with a short basal appendage. Petals unequal, the lowermost (the lip) usually larger, sometimes equal to or smaller than the other petals and with a short or long basal spur directed backwards. Stamens with very short free filaments and free or slightly coherent anthers; connective-appendages entire, generally oblong or ovate; lower 2 anthers each with an appendage extending into the spur and which excretes nectar. Ovary generally with numerous ovules; style usually deflexed downwards and thickened towards the tip. Fruit a loculicidal capsule with 3 contractile valves. Seeds usually smooth, with or without an aril.

A genus of over 400 species occurring throughout the world, but particularly in northern temperate regions.

Leaf-lamina base rounded or somewhat attenuate into the petiole;
    dwarf plants without trailing stems .   .   .   .   .   .3. *V. nannae*
Leaf-lamina base cordate; plants generally with trailing stems:
    Sepals without basal appendages; leaf-lamina ovate-cordate,
      pubescent above mainly between the lateral veins. . .1. *V. abyssinica*
    Sepals with short basal appendages; leaf-lamina reniform or
      suborbicular-cordate, pubescent above mainly along the
      lateral veins .   .   .   .   .2. *V. eminii*

1. **V. abyssinica** *Oliv.,* F.T.A. 1: 105 (1868); Engl., Hochgebirgsfl. Trop. Afr.: 308 (1892); W. Becker in E. & P. Pf., ed. 2, 21: 364, t. 159, fig. 34 (1925); W.F.K.: 10, fig. 10 (1948); F.P.N.A. 1: 631 (1948); F.W.T.A., ed. 2, 1: 107 (1954); N. Robson in F.Z. 1: 258, t. 41 (1960); Taton in F.C.B., Violaceae: 69 (1969); U.K.W.F.: 100, fig. (p. 98) (1974); Mountain Fl. S. Tanz.: 59, t. 10A (1982); Troupin, Fl. Rwanda 2: 434, fig. 137/1 (1983). Type: Ethiopia, *Schimper* 983 (K, lecto.!, BM, K, isolecto.)

Perennial herb; stems up to 60 cm. long, trailing or straggling through surrounding vegetation, often rooting at the nodes, glabrous or sparsely pubescent. Leaf-lamina broadly ovate, (9–)14–40 mm. long, (7–)12–33 mm. wide, generally longer than wide, the base cordate, the apex subobtuse to short acuminate, margin shallowly crenate-dentate, ± pubescent beneath, pubescent mainly between the lateral veins above; petiole 6–32 mm. long, pubescent or ± glabrous. Stipules foliaceous, ovate to lanceolate, 4–12 mm. long, laciniate, particularly in the lower half, pubescent or subglabrous. Flowers solitary, axillary, whitish, pale blue, purplish, mauve or violet-blue, the lip with darker striations; pedicels slender, (1–)2.8–7.2 cm. long, pubescent to glabrous, with a pair of linear-subulate bracteoles in the upper half, each 2–5 mm. long. Sepals subequal, lanceolate or linear-lanceolate, 5–7 mm. long, without basal appendages, generally hirsute along the midrib, sometimes with a ciliate margin. Petals unequal; upper 2 oblong to obovate, 7–9 mm. long, obtuse; lateral 2 ± oblanceolate, 7–9 mm. long, obtuse; lower petal (lip) elliptic, 5–7 mm. long, subobtuse, with a cylindric spur at the base, 2–4 mm. long, obtuse. Stamens ± one-third the length of the lateral petals, with orange or brownish connective appendages. Styles slightly exserted beyond the anthers. Capsule 6–7 mm. long, smooth, glabrous. Fig. 7/1–3.

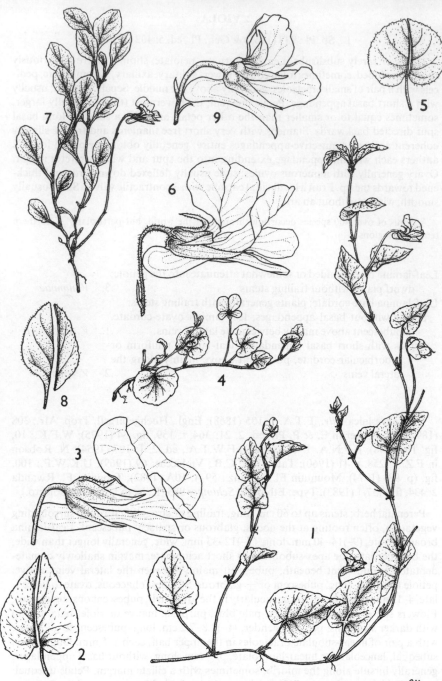

FIG. 7.   *VIOLA ABYSSINICA* – **1,** habit, x $\frac{2}{3}$; **2,** leaf, upper surface (right) and lower surface,
x 1; **3,** flower, x 3. *V. EMINII* – **4,** habit, x $\frac{2}{3}$; **5,** leaf, upper surface (right) and lower surface,
x 1; **6,** flower, x 3. *V. NANNAE* – **7,** habit, x $\frac{2}{3}$; **8,** leaf, upper surface (right) and lower surface,
x 1; **9,** flower, x 3. 1, from *Moore* 6A; 2, from *Geesteranus* 5636; 3, from *Drummond & Hemsley*
1545; 4, 6, from *Mabberley & McCall* 204; 5, from *Tweedie* 1280; 7–9, from *Greenway &
Kanuri* 14877. Drawn by Christine Grey-Wilson.

UGANDA.   Kigezi District: Mt. Mgahinga, Oct. 1929, *Snowden* 1569! & Virunga Mts., Mt. Mu-
havura, *Eggeling* 970!; Mbale District: NE. Elgon, July 1950, *Tweedie* 860!
KENYA.   Nakuru District: Mau Forest Reserve, Nov. 1971, *Magogo* 1493!; Londiani District:
Tinderet Forest Reserve, 6 km. SSE. of Timboroa Station, July 1949, *Maas Geesteranus* 5507;
Teita Hills, Yale Peak, Sept. 1953, *Drummond & Hemsley* 4314!
TANZANIA.   Lushoto District: Magamba Peak, Sept. 1945, *Greenway* 7543!; Morogoro District:
S. Uluguru Mts., Chenzema to Lukwangule Plateau, Jan. 1976, *Cribb & Grey-Wilson* 10447!;
Rungwe District: N. slopes Rungwe Mt., Feb. 1961, *Richards* 14272!
DISTR.   U 1–3; K 1, 3–7; T 2–4, 6–8; Nigeria, Cameroun, Fernando Po, Zaire, Ethiopia, Zambia,
Malawi, Zimbabwe, Mozambique, South Africa and Madagascar
HAB.   Upland forest margins, upland grassland, bushland and bamboo thickets;
1200–2700(–3350) m.

SYN.   *V. abyssinica* Oliv. var. *usambarensis* Fries in Acta Horti Berg. 8: 5 (1923). Types:
Tanzania, Lushoto District, Kwai, *Albers* 93, *Eick* 317, 401 & *Braun* in Herb. *Amani*
2852 and several other syntypes from W. Usambara Mts. (B, syn.†)

NOTE.   *V. abyssinica* is very variable, particularly in leaf size, general habit and in the amount
of pubescence. Plants growing in more exposed places tend to be rather smaller with more
rounded leaves and less dissected stipules and may be easily mistaken for *V. eminii,* however,
they can be recognised by the hairs on the upper surface of the leaves which are scattered
between, rather than along, the lateral veins. *Richards* 7672 from the Kitulo [Elton] Plateau
is typical of such plants which occur throughout the range of the species, generally growing
in grassy places rather than in forest.

1 x 2. V. abyssinica *Oliv.* x **V. eminii** *(Engl.) R.E. Fries;* R.E. Fries in Acta Horti
Berg. 8: 8 (1923); Grey-Wilson in K.B. 36: 125 (1981)

Plants intermediate in character. Leaf-lamina suborbicular to orbicular-ovate, up
to 30 mm. long and 30 mm. wide, pubescent or glabrous beneath, pubescent on the
main lateral veins above and also usually between the veins, especially towards the
margin of the leaf. Sepals usually with very short basal appendages which are often
rather difficult to observe.

UGANDA.   Toro District: Ruwenzori Mts., Bujuku valley, Aug. 1933, *Eggeling* 1264!; Kigezi
District: Mt. Mgahinga, Oct. 1947, *Purseglove* 2520! & Mt. Muhavura, *Hedberg* 2260!
KENYA.   Trans-Nzoia District: Kitale, Nov. 1940, *Bally* 1258!; Aberdare Mts., Mar. 1922, *Fries*
2549!
TANZANIA.   Mbulu District: Hanang, Dec. 1929, *B.D. Burtt* 2254!; Moshi District: Kilimanjaro,
Bismark Hill, Feb. 1934, *Greenway* 3868!; Morogoro District: Uluguru Mts., Lukwangule
Plateau, Mar. 1955, *Semsei* 2051!
DISTR.   U 2, 3; K 3, 4; T 2, 6; Zaire
HAB.   Upland evergreen forest, upland grassland and bamboo thickets; 2500–3450 m.

NOTE.   A widespread and variable hybrid. Forms from Mt. Meru and Kilimanjaro with large
suborbicular-cordate leaves up to 30 mm. long and wide appear to represent a complex hybrid
origin perhaps with introgression towards *V. eminii; B.D. Burtt* 4104, *Geilinger* s.n., 7 Dec.
1932, and *Richards* 25616 are typical of such plants.

2. V. eminii *(Engl.) R.E. Fries* in Acta Horti Berg. 8: 6, t. 1/3 (1923); W.F.K: 10
(1948); F.P.N.A. 1: 632 (1948); A.V.P.: 132 (1957); Taton in F.C.B., Violaceae: 71
(1969); U.K.W.F.: 100, fig. on p. 98 (1974); Troupin, Fl. Rwanda 2: 434, fig. 137/2
(1983). Type: Uganda, Ruwenzori Mts., *Stuhlmann* 2416 (B, holo.†) – see note below

Perennial herb; stems usually trailing, up to 40 cm. long though generally less, often
rooting at the nodes, sparsely to usually densely pubescent, especially when young.
Leaf-lamina reniform to suborbicular, 6–22 mm. long, 6–24 mm. wide, often wider
than long, the base cordate, the apex obtuse, rarely slightly apiculate, margin shallowly

crenate, usually pubescent beneath especially along the midrib and lateral veins, pubescent above, the hairs confined to the midrib, lateral veins and sometimes the marginal area; petiole 4–10 mm. long, usually pubescent. Stipules ovate or lanceolate, 3–8 mm. long, laciniate. Flowers solitary, axillary, violet, purple-blue or mauve, rarely white, the lip with darker striations; pedicels slender, 2.4–7 cm. long, pubescent to subglabrous, with a pair of linear-lanceolate bracteoles in the upper half, each 4–6 mm. long. Sepals ± equal, triangular-lanceolate, 6–7 mm. long, with a short basal appendage, usually pubescent along the midrib, the margin ± ciliate. Petals unequal; upper 2 obovate, 7–10(–12) mm. long; lateral 2 obovate-oblong to oblanceolate, 8–10(–12) mm. long; lower petal (lip) elliptic-oblong, (5.5–)6–8 mm. long, subobtuse, with a cylindric obtuse spur, 2.5–3 mm. long. Stamens ± one-third the length of the lateral petals, each with a non-decurrent connective appendage. Style ± 2 mm. long, slightly exserted beyond the anthers. Capsule 5–7 mm. long, smooth, glabrous. Fig. 7/4–6.

UGANDA.   Karamoja District: Mt. Debasien, 1936, *Eggeling* 2707!; Mbale District: Elgon, Bulambuli, Nov. 1933, *Thomson* 2253! & above Budadiri [Butadiri], Mudangi track, Dec. 1967, *Hedberg* 4556!
KENYA.   Trans-Nzoia District: NE. Elgon, Dec. 1954, *Tweedie* 1280!; Elgeyo District: near Kamalagon [Kameligon], Aug. 1969, *Mabberley & McCall* 204!; S. Nyeri District: Mt. Kenya, above Kiganjo, Dec. 1959, *A. Moore* 4!
TANZANIA.   Masai District: Ngorongoro, Embagai [Empakaai] Crater, Feb. 1973, *Frame* P15!; Moshi District: Kilimanjaro, Dec. 1932, *Geilinger* 4290! & Shira Plateau, Feb. 1928, *Haarer* 1125
DISTR.   U 1–3; K 3–5; T 2, 6; Zaire, Rwanda, Burundi
HAB.   Upland forest fringes, upland grassland and moor, occasionally in bamboo thickets; 2150–4050 m.

SYN.   *V. abyssinica* Oliv. var. *eminii* Engl., P.O.A. C: 276 (1895); De Wild., Pl. Bequaert. 1: 246 (1922)
       V. abyssinica Oliv. var. *ulugurensis* Engl. in E.J. 28: 437 (1900). Types: Tanzania, Uluguru Mts. Lukwangule Plateau, *Goetze* 313 & *Stuhlmann* 9155 (B, syn.†)
       V. mearnsii Standl. in Smithson. Misc. coll. 68(5): 9 (1917), *non* Merrill. Type: Kenya, Mt. Kenya, 3000 m. *Mearns* 1718 (US, holo.)

NOTE.   The type specimen of *V. eminii* has been destroyed. Hedberg, A.V.P.: 132 (1957), designates *Hedberg* 434 as a neotype, however I do not agree with this. The specimen is not typical of *V. eminii* and like Tennant I think that it probably represents a plant of hybrid origin, *V. abyssinica* x *V. eminii*.

3. V. nannae R.E. *Fries* in Acta Horti Berg. 8: 10, t. 1/5 (1923). Type: Kenya, Aberdare Mts., Kinangop peak, *Fries* 2695 (UPS, holo., K, iso.!)

Dwarf perennial herb usually 2–6 cm. tall, rarely to 15 cm.; stems usually glabrous. Leaf-lamina ovate to suborbicular, 5–20 mm. long, 4–17 mm. wide, generally rather longer than wide, the base rounded or somewhat attenuate into the petiole, the apex obtuse, glabrous beneath, but generally pubescent along and close to the margin above, margin shallowly crenate with 3–4 crenations along each side, or subentire. Stipules linear-lanceolate to oblong-lanceolate, 3.5–6 mm. long, acute, entire or subentire, margin ciliate. Flowers solitary, axillary, violet or violet-purple; pedicels slender, 2.6–12.5 cm. long, glabrous or pubescent towards the top, with a pair of subopposite linear-lanceolate bracteoles in the upper half, each 2–4 mm. long. Sepals ± equal, narrowly triangular-lanceolate, 4–5 mm. long, with a very short basal appendage, pubescent towards the base or along the midrib, usually with a ciliate margin. Petals unequal; upper 2 oval-obovate, 6.5–8.5 mm. long; lateral 2 oval-oblanceolate, 6.5–8 mm. long; lower petal (lip) elliptic, 4–6 mm. long, with a cylindric obtuse spur,

± 2.5 mm. long. Stamens ± one-third the length of the lateral petals, each with a non-decurrent orange-brown connective-appendage. Capsule 4–5 mm. long, smooth, glabrous. Fig. 7/7–9.

KENYA. Nakuru District: Mau Summit, 1930, *Mettam* 181!; Naivasha District: South Kinangop, Nov. 1953, *Verdcourt* 1049!; Kisumu-Londiani District: Mt. Blackett, July 1933, *C.G. Rogers* 521! in part; Masai District: Nasampolai valley, June 1971, *Greenway & Kanuri* 14877! in part

DISTR. K 3, ?4, 5, 6; not known elsewhere

HAB. Upland grassland; 2550–3300 m.

NOTE. Hybrids between *V. nannae* and *V. eminii* probably occur where the two species grow close together (see Grey-Wilson in K.B. 36: 125 (1981)). *C.G. Rogers* 521 and *Greenway & Kanuri* 14877 are both mixed gatherings with some specimens of *V. nannae* and some apparently of such hybrid origin. However, some doubt must remain as these hybrids can look very similar to individuals of *V. abyssinica* which are found in exposed grassy places (see p. 27), though they generally have rather rounded or ± truncate leaf bases, very little leaf pubescence and entire or almost entire stipules. This problem needs to be investigated in 'the field'. *Viola duriprati* R.E. Fries in Acta Horti Berg. 8: 8, t. 1/1–2 (1923), based on Kenya, Aberdare Mts., *Fries* 386 (UPS, holo., K, iso.!), appears to be of more complicated origins, possibly *V. abyssinica* x *V. nannae;* the leaves are pubescent on both surfaces and have a subtruncate base to the lamina and the stipules are subentire. *Chandler* 2226a, also from the Aberdares, is very similar to the last mentioned. *Viola duriprati* has been made a variety of *V. eminii,* var. *duriprati* (R.E. Fries) Stork.

3. HYBANTHUS

Jacq., Enum. Syst. Pl. Ins. Carib.: 2, 17 (1760)

Ionidium Vent., Jard. Malm. 1, sub. t. 27 (1803)

Annual or perennial herbs, subshrubs or occasionally shrubs with simple or much-branched stems. Leaves spirally arranged, occasionally opposite or fasciculate, petiolate or sessile, the lamina simple, entire or toothed; stipules small and persistent, rarely foliaceous. Flowers zygomorphic, solitary in the leaf-axils, rarely terminal or in clusters; pedicels articulate in the middle or the upper half, with an opposite or suboppose pair of bracteoles. Sepals subequal, without a basal appendage. Petals unequal, the lowermost one (anterior) usually much larger than the others and clawed, with a short basal spur. Stamens unequal, free or somewhat coherent, the lower pair each with a short nectar-secreting appendage which extends into the spur; anthers with a thin dorsal connective-appendage. Fruit a loculicidal capsule with 3 contractile valves. Seeds pitted or striate, usually with a small aril.

A genus of about 150 species distributed throughout the tropics of the Old and New Worlds.

Leaves spirally arranged even on the older stems, linear to ovate,
    margins entire or subentire (except in K 7, T 3, 7) .   .   .1. *H. enneaspermus*
Leaves in distinct fascicles on the old stems, ovate to oblanceolate,
    margin clearly serrate or dentate (K 4) .   .   .   .   .   .2. *H. fasciculatus*

1. **H. enneaspermus** *(L.) F. Muell.,* Fragm. Phytogr. Austral. 10: 81 (1876) & Census Austr. Pl. 1: 6 (1882); Exell & Mendonça, C.F.A. 1: 76 (1937); F.P.N.A. 1: 630 (1948); F.W.T.A., ed. 2, 1: 106, fig. 33 (1954); N. Robson in Bol. Soc. Brot., sér. 2A, 32: 167 (1958) & in F.Z. 1: 254, t. 40 (1960); Tennant in K.B. 16: 431 (1963); Agnew,

U.K.W.F.: 100 (1974); Grey-Wilson in K.B. 36: 103 (1981) & 39: 771 (1984). Type: Ceylon, *Hermann Herbarium* 1. 19 (BM, lecto.!)

Annual or perennial herb to 60 cm. tall, more rarely subshrubby to 3 m.; stems simple or moderately branched, subglabrous to densely pubescent, often becoming glabrescent, ± terete or ridged. Leaves sessile to subsessile; lamina linear to linear-lanceolate, oblanceolate or elliptic-lanceolate, 0.5–8 cm. long, 0.1–1.4 cm. wide, the base cuneate, the apex acute to subobtuse, glabrous to densely pubescent, hirsute or scabrid, margin subentire to crenate-serrate or dentate-serrate, lateral veins 4–7 pairs; stipules linear-lanceolate to subulate, 1.5–4 mm. long, margin ciliate usually. Flowers solitary, axillary, pink, reddish violet, mauve, blue, orange-red, white or bicoloured; pedicels slender, 6–21 mm. long, with a pair of small subulate bracteoles, up to 2 mm. long, in the upper half, glabrous or pubescent. Sepals lanceolate, subequal, 2.5–4 mm. long, acute, glabrous or pubescent. Petals very unequal; upper 2 elliptical, 3–5.5 mm. long, symmetric, acuminate; lateral 2 triangular-oblong, 4–6 mm. long, often rather expanded at the apex and obtuse; lower petal (lip) suborbicular to subcordate, 8–19 mm. long, with a claw exceeding the calyx and including a short obtuse spur. Stamens much shorter than the lateral petals, the lower 2 with longer filaments which bear pilose spur-appendages; anthers each with a short ovate connective-appendage. Ovary globose, glabrous; style slightly exceeding the anthers and thickened towards the tip. Capsule 5–8 mm. long, 3-lobed, glabrous. Seeds ovoid-ellipsoidal, ± 1.5 mm. long, longitudinally ribbed.

Leaves with prominent reticulate venation on the upper surface;
    lamina oblanceolate, increasing in size up the stem . . . .e. var. **nyassensis**
Leaves without prominent reticulate venation on the upper surface; lamina linear to lanceolate, elliptic-lanceolate or elliptic, generally decreasing in size up the stem:
    Stems branched in the upper half; leaves of lateral branches
      markedly smaller than those of the main stem . . . .b. var. **diversi-folius**
    Stems branched in the lower half (except var. *latifolius* which is the only shrubby var.); leaves of lateral branches not markedly smaller than those of the main stem:
      Stems densely white-hirsute . . . . .c. var. **tsavoensis**
      Stems subglabrous to pubescent or ± scabrid:
        Pedicels 17–26 mm. long:
          Leaf-lamina narrowly lanceolate-elliptic, long acuminate; leaf-margin obscurely toothed . . . .f. var. **pseudo-caffer**
          Leaf-lamina lanceolate to lanceolate-elliptic, shortly acuminate; leaf-margin incise-dentate. . . .d. var. **pseudo-danguyanus**
        Pedicels 5–16 mm. long:
          Leaves densely congested, especially along the lateral shoots, rarely more than 3 cm. long . . . .h. var. **densifolius**
          Leaves not densely congested, often more than 3 cm. long:
            Leaves linear to narrowly elliptic, rarely lanceolate or lanceolate-elliptic, 1–10 mm. wide, obscurely toothed to ± entire; annuals or short-lived perennials of dry bushland, rarely more than 30 cm.

FIG. 8. *HYBANTHUS ENNEASPERMUS* var. *ENNEASPERMUS – Group 1:* **1,** habit, x ⅔; **2,** leaf, upper surface, x 1. Group 2: **3,** habit, x ⅔; **4,** leaf, upper surface, x 1; **5,** flower, lateral view, x 3; **6,** flower, view from above, x 3; **7,** fruit, x 3; **8,** seed, front and lateral view, x 4. *H. ENNEASPERMUS* var. *DIVERSIFOLIUS* – **9,** habit, x ⅔; **10,** leaf, upper surface, x 1; **11,** leaf, lower surface, x 1. 1, 2, from *Bullock* 2492; 3, 4, 7, 8, from *Greenway & Kanuri* 13571; 5, 6, from *Drummond & Hemsley* 4042; 9–11, from *R.M. Graham* in *F.D.* 1983. Drawn by Christine Grey-Wilson.

tall .   .   .   .   .a.   var. **enneasper-**
                                                          mus

Leaves lanceolate to lanceolate-elliptic, acuminate,
6–14 mm. wide, shallowly toothed; shrub or
subshrub of evergreen forests, 90–120 cm. tall   .g.   var. **latifolius**

a. var. **enneaspermus**

Annual or perennial herb to 30 cm. tall, rarely more; stem simple or moderately branched in the lower half, erect or decumbent, glabrous to pubescent or glabrescent. Leaf-lamina linear to linear-lanceolate or linear-elliptic, 10–76 mm. long, 1–10 mm. wide. Flowers pink, magenta, reddish violet, white or bicoloured; pedicels 5–8 mm. long. Lower petal (lip) 8–15 mm. long. Fig. 8/1–8.

UGANDA.   Karamoja District: Oropoi, June–July 1930, *Liebenberg* 145!; Teso District: Bugondo, Dec. 1931, *Chandler* 374! & Serere, June 1932, *Chandler* 675!

KENYA.   Masai District: Namanga, Dec. 1963, *Verdcourt* 3845!; Teita District: Voi Gate West, Dec. 1966, *Greenway & Kanuri* 12720!; Kwale District: Shimba Forest, Mar. 1968, *Magogo & Glover* 391! & between Samburu and Mackinnon Road, Aug. 1953, *Drummond & Hemsley* 4042!

TANZANIA.   Tanga District: Machui, Jan. 1956, *Faulkner* 1792!; Ufipa District: Milepa, Feb. 1950, *Bullock* 2492!; Iringa District: between Izazi and Kilindimo, Feb. 1962, *Polhill & Paulo* 1331!; Zanzibar I., Mazizini [Massazine], Feb. 1961, *Faulkner* 2765!

DISTR.   U 1, 3; K 1, 4–7; T 1–7; Z; widespread in Africa, also in Asia

HAB.   Lowland evergreen forest, wooded grassland, grassland, cultivated and waste places; sea-level to 1250 m.

SYN.   *Viola enneasperma* L., Sp. Pl.: 937 (1753); Willd., Sp. Pl. 1: 1171 (1798)
        V. suffruticosa L., Sp. Pl.: 937 (1753); Willd., Sp. Pl. 1: 1171 (1798). Type: Ceylon, *Hermann Herbarium* 1. 41 (BM, holo.!)
        Ionidium enneaspermum (L.) Vent., Jard. Malm., sub. t. 27 (1803); Oliv., F.T.A. 1: 105 (1868); Engl., P.O.A. C: 277 (1895)
        I. suffruticosum (L.) Roem. & Schultes, Syst. Veg. 5: 394 (1819)

NOTE.   Var. *enneaspermus* is a very variable and widespread variety. Two main forms can be informally recognised, though intermediates also occur. The first group (fig. 8/1, 2) has lanceolate-elliptic leaves 4–11 mm. wide and is confined mainly to K 6–7, T 3–7, though it is to be found in other parts of Africa as well as Asia. Indeed this group closely matches the type material in the Hermann Herbarium. The second group (fig. 8/3–8) has linear or linear-lanceolate leaves 1–3(–4) mm. wide and is found over much of tropical Africa including U 1, 3, K 1, 4–7, T 1–3, 6, 7 and Z.

There is a large synonymy involved here both within Africa and in India and SE. Asia. Readers are best referred to Tennant in K.B. 16: 431 (1969), though it should be taken into account that he took a very broad view of *H. enneaspermus* and some of the 'synonyms' certainly deserve recognition as separate taxa.

b. var. **diversifolius** Grey-Wilson in K.B. 36: 106 (1981). Type: Kenya, Kilifi District, Arabuko, *R.M. Graham* in *F.D.* 1983 (K, holo.!, iso.!)

Perennial herb to 40 cm. tall; stems erect or decumbent, branched mainly in the upper half, finely pubescent to subglabrous when young. Leaf-lamina linear-elliptic to elliptic, 5–65 mm. long, 2–12 mm. wide, those of the lateral branches markedly smaller than those of the main stem. Pedicels 8–11 mm. long. Lower petal (lip) 10–15 mm. long. Fig. 8/9, 10.

KENYA.   Kwale District: Kwale, *R.M. Graham* in *F.D.* 1944! & Shimba Hills, Lango ya Mwagandi [Longomwagandi] area, Nov. 1968, *Magogo & Estes* 1185!; Kilifi District: Sokoke, Aug. 1945, *Jeffery* 298!

DISTR.   K 7; not known elsewhere

HAB.   Grassland; 50–250 m.

FIG. 9. *HYBANTHUS ENNEASPERMUS* var. *NYASSENSIS* – **1,** habit, x $\frac{2}{3}$; **2,** leaf, upper surface, x 1. *H. ENNEASPERMUS* var. *PSEUDOCAFFER* – **3,** habit, x $\frac{2}{3}$; **4,** leaf, upper surface, x 1. *H. ENNEASPERMUS* var. *LATIFOLIUS* – **5,** habit, x $\frac{2}{3}$; **6,** leaf, upper surface, x 1. 1, 2, from *Milne-Redhead & Taylor* 8268; 3, 4, from *Faulkner* 2026; 5, 6, from *Richards* 6769. Drawn by Christine Grey-Wilson.

Note.   *Faden* 74/1216 comes midway between var. *diversifolius* and var. *tsavoensis* in leaf characters.

c. var. **tsavoensis** *Grey-Wilson* in K.B. 36: 106 (1981). Type: Kenya, Tsavo Park, Sala Gate to Sobo Rocks, *Greenway & Kanuri* 12916 (K, holo.!)

Perennial herb to 24 cm. tall; stems branched from the base, otherwise simple, ascending, densely white-hirsute. Leaf-lamina broadly elliptic-obovate, 20–52 mm. long, 9–16 mm. wide, decreasing in size up the stem. Flowers orange-red; pedicels 11–21 mm. long. Lower petal (lip) 15–17 mm. long. Fig. 10/1, 2.

Kenya.   Teita District: Tsavo Park, Sala Gate to Sobo Rocks, Dec. 1966, *Greenway & Kanuri* 12916!
Distr.   K 7; not known elsewhere
Hab.   Sandstone rocks; ± 240 m.

d. var. **pseudodanguyanus** *Grey-Wilson*, in K.B. 39: 771 (1984). Type: Kenya, Meru National Park, Mughwango Plains, *P.H. Hamilton* 203 (EA, holo.!)

Perennial herb to 90 cm. tall, though often less, branched towards the base, otherwise ± simple; stems covered in a fine pubescence of curved hairs. Leaf-lamina lanceolate to elliptic-lanceolate, 1.3–6.7 x 0.5–2.5 cm., generally increasing in size up the stem, finely pubescent above and beneath, or only beneath, the hairs confined to the midrib and main lateral veins; margin incised-dentate. Flowers pink or salmon; pedicels 14–23 mm. long. Lower petal (lip) 13–15 mm. long.

Kenya.   Northern Frontier Province: Benani [Benane], Jan. 1956, *J. Adamson* 588!; Meru District: Meru National Park, near Leopard Rock Camp, Mar. 1968, *Mwangangi & Fosberg* 617/B! & Mughwango Plains, Dec. 1977, *P.H. Hamilton* 203!
Distr.   K 1, 4; not known elsewhere
Hab.   Rocky slopes and lowland bushland; ± 580 m.

e. var. **nyassensis** *(Engl.) N. Robson* in Bol. Soc. Brot., sér. 2A, 32: 168 (1958) & in F.Z. 1: 257, t. 40/B (1960); Grey-Wilson in K.B. 36: 109 (1981). Lectotype, chosen here: Malawi, Shire Highlands, Blantyre, *Last* (K, lecto.!)

Perennial herb to 30 cm. tall, though often less; stems erect, branched from the base otherwise simple, often reddish, usually scabrid. Leaf-lamina oblanceolate, 16–44 mm. long, 5–8 mm. wide, increasing in size up the stem, with prominent reticulate venation above. Flowers pink, purplish, lilac or violet; pedicels 11–18 mm. long. Lower petal (lip) 12–20 mm. long. Fig. 9/1, 2.

Tanzania.   Songea District: 9 km. W. of Songea, Jan. 1956, *Milne-Redhead & Taylor* 8268! & Kigonsera, Dec. 1973, *Mhoro* 1697! & Mangua, *Busse* 1316!
Distr.   T 8; N. Mozambique, Malawi, N. Zambia
Hab.   Sandy soil, often in *Brachystegia-Uapaca* woodland; ± 960 m.

Syn.   *Ionidium nyassense* Engl., P.O.A. C: 277 (1895)
         Hybanthus nyassensis (Engl.) Engl. in E.J. 55: 400 (1919)

f. var. **pseudocaffer** *Grey-Wilson* in K.B. 36: 109 (1981). Type: Tanzania, Tanga District, Steinbruch Forest, *Faulkner* 2026 (K, holo.!, iso.!)

Perennial herb to 60 cm. tall; stems erect or decumbent, moderately branched, usually finely pubescent. Leaf-lamina lanceolate-elliptic, acute, 26–85 mm. long, 5–13 mm. wide. Flowers reddish purple; pedicels 17–26 mm. long. Lower petal (lip) 12–15 mm. long. Fig. 9/3, 4.

Tanzania.   Tanga District: Steinbruch Forest, July 1957, *Faulkner* 2026!
Distr.   T 3; not known elsewhere
Hab.   Lowland evergreen forest, in grassy places; ± 65 m.

g. var. **latifolius** *(De Wild.) Engl.* in E.J. 55: 398 (1919); N. Robson in Bol. Soc. Brot., sér. 2A, 32: 170 (1958) & in F.Z. 1: 258, t. 40/D (1960); Grey-Wilson in K.B. 36: 109 (1981). Type: Zaire, Oubangui, R. Mongala, Abumombazi, *Thonner* 203 (BR, holo.)

FIG. 10. *HYBANTHUS ENNEASPERMUS* var. *TSAVOENSIS* – **1**, habit, x $\frac{2}{3}$; **2**, leaf, upper surface, x 1. *H. FASCICULATUS* – **3**, habit, x $\frac{2}{3}$; **4**, leaf, upper surface, x 1; **5**, flower, lateral view, x 3; **6**, flower, lateral half-section x 3. 1, 2, from *Greenway & Kanuri* 12916; 3, 4, 6, from *Hamilton* 163; 5, from *Ament & Magogo* 278. Drawn by Christine Grey-Wilson.

Shrub or subshrub of loose rather spindly habit to 3 m. tall, though often less; stems moderately branched, slightly hispid to glabrescent or glabrous. Leaf-lamina elliptic-lanceolate, 20–48 mm. long, 6–14 mm. wide, acuminate. Flowers mauve or purplish; pedicels 7–16 mm. long. Lower petal (lip) 14–15 mm. long. Fig. 9/5, 6.

TANZANIA.   Rungwe District: Rungwe Forest, Aug. 1924, *Stolz* 866! & Rungwe Mt., Oct. 1956, *Richards* 6769!
DISTR.   **T** 7; Mozambique, Zaire and parts of W. Africa
HAB.   Evergreen forest; ± 2400 m.

SYN.   *Ionidium enneaspermum* (L.) Vent. var. *latifolium* De Wild., Pl. Thonn. Cong., sér. 2: 238, t. 17 (1911)

h. var. **densifolius** *Grey-Wilson* in K.B. 39: 772 (1984). Type: Kenya, Meru National Park, Golo, *P.H. Hamilton* 500 (EA, holo.!)

Thin-stemmed shrub (height unknown), branched, the uppermost branches short and rather congested; stems finely pubescent with short curved hairs, becoming ± glabrescent eventually. Leaf-lamina narrowly elliptical, 5–35 x 1.5–5 mm., those of the main shoots rather longer than those of lateral shoots, all somewhat congested, finely pubescent above and beneath; margin entire or indistinctly toothed. Flowers pinkish mauve; pedicels 6–8 mm. long. Lower lip 5–7 mm. long.

KENYA.   Meru District: Meru National Park, Golo, May 1979, *P.H. Hamilton* 500!
DISTR.   **K** 4; not known elsewhere
HAB.   *Acacia* wooded grassland; ± 550 m.

2. H. fasciculatus *Grey-Wilson* in K.B. 36: 110, fig. 3C–F (1981). Type: Kenya, Meru National Park, Rainkombe kopje, Nov. 1977, *P.H. Hamilton* 163 (K, holo.!, EA, iso.)

Low shrub to 0.8–2 m. tall; stems fairly densely branched, puberulous at first but soon becoming glabrescent and woody, to 5 mm. diameter, the bark rather corky and whitish. Leaves alternate on the young shoots but in distinct lateral clusters on the older stems, sessile or with a very short petiole up to 0.5 mm. long; lamina oblanceolate to obovate or elliptic-oblanceolate, 1–3.1 cm. long, 0.5–1.3 cm. wide, the base cuneate, the apex subobtuse, scabrid or finely pubescent above, but becoming glabrescent, glabrous beneath, margin serrulate, lateral veins 3–5 pairs; stipules linear-lanceolate, 2.5–3 mm. long, ciliate margined. Flowers solitary, axillary, whitish with a bright pink lip; pedicels 18–28 mm. long, puberulous; bracteoles 2, subulate, 1–1.5 mm. long, borne on the upper half of the pedicel. Sepals linear-lanceolate, 3.5–4.5 mm. long, acute, subequal. Petals very unequal; upper 2 oblong, ± 4.5 mm. long, shortly apiculate; lateral 2 ovate-triangular, 5–5.5 mm. long, somewhat broadened at the apex, obtuse; lower petal (lip) suborbicular to broadly oval, 14–20 mm. long, 8–11 mm. wide, abruptly narrowed towards the base into a claw which is slightly longer than the calyx. Stamens ± half the length of the lateral petals; anthers with a broad oval connective-appendage. Capsule ± 7 mm. long, 3-lobed, glabrous. Fig. 10/3–6.

KENYA.   Meru District: Mughwango, May 1972, *Ament & Magogo* 278! & Nov. 1980, *P.H. Hamilton* 787! & Rainkombe Kopje, Nov. 1977, *P.H. Hamilton* 163! & Dec. 1977, 199!
DISTR.   **K** 4; not known elsewhere
HAB.   Steep rocky hillslopes, lowland bushland; 580–700 m.

NOTE.   This species has been called *H. danguyanus* H. Perrier which is native to Madagascar, however, the Madagascan plant is very different – an annual with strictly alternate leaves, shorter pedicels and smaller flowers.

INDEX TO VIOLACEAE

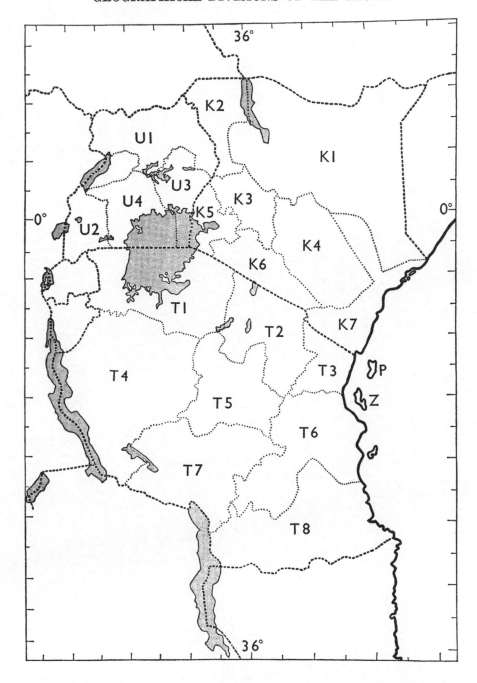

T - #0684 - 101024 - C0 - 244/170/5 - PB - 9789061913306 - Gloss Lamination